Digital Image Processing for Ophthalmology

Detection of the Optic Nerve Head

Digital Image Processing for Ophthalmology: Detection of the Optic Nerve Head
Xiaolu Zhu, Rangaraj M. Rangayyan, and Anna L. Ells

ISBN-13: 978-3-031-00521-3   paperback
ISBN-13: 978-3-031-01649-3   ebook

DOI 10.1007/978-3-031-01649-3

A Publication in the Springer series
SYNTHESIS LECTURES ON BIOMEDICAL ENGINEERING

Lecture #40
Series Editor: John D. Enderle, *University of Connecticut*
Series ISSN
Synthesis Lectures on Biomedical Engineering
Print 1930-0328   Electronic 1930-0336

Synthesis Lectures on Biomedical Engineering

Editor
John D. Enderle, *University of Connecticut*

Lectures in Biomedical Engineering will be comprised of 75- to 150-page publications on advanced and state-of-the-art topics that spans the field of biomedical engineering, from the atom and molecule to large diagnostic equipment. Each lecture covers, for that topic, the fundamental principles in a unified manner, develops underlying concepts needed for sequential material, and progresses to more advanced topics. Computer software and multimedia, when appropriate and available, is included for simulation, computation, visualization and design. The authors selected to write the lectures are leading experts on the subject who have extensive background in theory, application and design.

The series is designed to meet the demands of the 21st century technology and the rapid advancements in the all-encompassing field of biomedical engineering that includes biochemical, biomaterials, biomechanics, bioinstrumentation, physiological modeling, biosignal processing, bioinformatics, biocomplexity, medical and molecular imaging, rehabilitation engineering, biomimetic nano-electrokinetics, biosensors, biotechnology, clinical engineering, biomedical devices, drug discovery and delivery systems, tissue engineering, proteomics, functional genomics, molecular and cellular engineering.

Digital Image Processing for Ophthalmology: Detection of the Optic Nerve Head
Xiaolu Zhu, Rangaraj M. Rangayyan, and Anna L. Ells
2011

Modeling and Analysis of Shape with Applications in Computer-Aided Diagnosis of Breast Cancer
Denise Guliato and Rangaraj M. Rangayyan
January 2011

Analysis of Oriented Texture with Applications to the Detection of Architectural Distortion in Mammograms
Fábio J. Ayres, Rangaraj M. Rangayyan, and J. E. Leo Desautels
2010

Fundamentals of Biomedical Transport Processes
Gerald E. Miller
2010

Digital Image Processing for Ophthalmology

Detection of the Optic Nerve Head

Xiaolu Zhu
University of Calgary, Calgary, Alberta, Canada

Rangaraj M. Rangayyan
University of Calgary, Calgary, Alberta, Canada

Anna L. Ells
Alberta Children's Hospital, Calgary, Alberta, Canada

SYNTHESIS LECTURES ON BIOMEDICAL ENGINEERING #40

ABSTRACT

Fundus images of the retina are color images of the eye taken by specially designed digital cameras. Ophthalmologists rely on fundus images to diagnose various diseases that affect the eye, such as diabetic retinopathy and retinopathy of prematurity.

A crucial preliminary step in the analysis of retinal images is the identification and localization of important anatomical structures, such as the optic nerve head (ONH), the macula, and the major vascular arcades. Identification of the ONH is an important initial step in the detection and analysis of the anatomical structures and pathological features in the retina. Different types of retinal pathology may be detected and analyzed via the application of appropriately designed techniques of digital image processing and pattern recognition. Computer-aided analysis of retinal images has the potential to facilitate quantitative and objective analysis of retinal lesions and abnormalities. Accurate identification and localization of retinal features and lesions could contribute to improved diagnosis, treatment, and management of retinopathy.

This book presents an introduction to diagnostic imaging of the retina and an overview of image processing techniques for ophthalmology. In particular, digital image processing algorithms and pattern analysis techniques for the detection of the ONH are described. In fundus images, the ONH usually appears as a bright region, white or yellow in color, and is indicated as the convergent area of the network of blood vessels. Use of the geometrical and intensity characteristics of the ONH, as well as the property that the ONH represents the location of entrance of the blood vessels and the optic nerve into the retina, is demonstrated in developing the methods.

The image processing techniques described in the book include morphological filters for preprocessing fundus images, filters for edge detection, the Hough transform for the detection of lines and circles, Gabor filters to detect the blood vessels, and phase portrait analysis for the detection of convergent or node-like patterns. Illustrations of application of the methods to fundus images from two publicly available databases are presented, in terms of locating the center and the boundary of the ONH. Methods for quantitative evaluation of the results of detection of the ONH using measures of overlap and free-response receiver operating characteristics are also described.

KEYWORDS

digital image processing, detection of blood vessels, detection of circles, edge detection, Gabor filter, Hough transform, morphological filters, node maps, optic disc, optic nerve head, pattern recognition, phase portraits, retinal fundus images

Xiaolu dedicates this book to her parents:
Qingqiao Wang *and* Guoyou Zhu

Contents

Preface

Large numbers of fundus images of the retina are being analyzed by ophthalmologists around the world. Digital imaging techniques and computing resources have been improving at rapid rates and finding more and more pratical applications. Over the past 20 years, researchers have been applying digital image processing techniques to ophthalmology with the aim of improved diagnosis.

This is an introductory book on digital image processing for application in ophthalmology. In particular, methods for automated detection of landmark features in retinal fundus images, such as the optic nerve head, fovea, and blood vessels are described in detail. The digital image processing techniques described include morphological filters, edge detectors, Gabor filters, the Hough transform, and phase portrait analysis. Methods for quantitative evaluation of the results of detection, such as distance measures and free-response receiver operating characteristics are also described. The image processing and quantitative evaluation techniques are not limited to medical applications.

The methods described in this book are mathematical in nature. It is assumed that the reader is proficient in advanced mathematics and familiar with basic notions of data, signal, and image processing. The methods of image modeling and analysis are suitable for inclusion in courses for students in the final year of bachelor's programs in electrical engineering, computer engineering, mathematics, physics, computer science, biomedical engineering, and bioinformatics. The techniques should also be useful to researchers in various areas of image modeling and analysis, and could be included in graduate courses on digital image processing, medical imaging, and related topics. The book is copiously illustrated with images and examples of application to facilitate efficient comprehension of the notions and methods presented.

We wish our readers success in their studies and research.

Xiaolu Zhu
Rangaraj M. Rangayyan
Anna L. Ells

Calgary, Alberta, Canada
January, 2011.

Acknowledgments

The research work described in this book was supported by the Natural Sciences and Engineering Research Council of Canada.

We thank Dr. Fábio J. Ayres for his assistance in parts of the research work described in this book.

Xiaolu Zhu
Rangaraj M. Rangayyan
Anna L. Ells

Calgary, Alberta, Canada
January, 2011.

CHAPTER 1

Introduction

1.1 DIAGNOSTIC IMAGING OF THE EYE

1.1.1 ANATOMY AND PHYSIOLOGY OF THE EYE

A major portion of the information acquired and processed by a human being is through the eyes and the visual system. The eye consists of three concentric layers [1]. The outermost fibrous sclera is opaque and accounts for five sixths of the globe of the eye. The sclera continues anteriorly into the transparent cornea. The middle vascular coat or uveal tract is made up of the choroid, ciliary body, and iris. The choroid is a highly vascularized structure, accounting for 80% the total ocular blood flow. The innermost layer is the light-sensitive retina, where light energy, focused by the lens, is transformed into neural signals, which are transmitted to the brain along the optic nerve through the optic nerve head (ONH) or optic disc. The central retinal vein and artery enter the eye within the trunk of the optic nerve through the ONH.

The retina is a thin layer of neural cells that lines the back of the eyeball of vertebrates and some cephalopods. In vertebrate embryonic development, the retina and the optic nerve originate from outgrowths of the developing brain. Hence, the retina is a part of the central nervous system. The retina is the only part of the central nervous system that can be imaged directly [2, 3]. The retina offers relatively easy assessment and early diagnosis of not only local ocular conditions but also many systemic diseases, including diabetes, arteriosclerosis, hypertension, and retinopathy of prematurity (ROP), which would otherwise require complicated, expensive, and time-consuming investigations.

1.1.2 DISEASES AND CONDITIONS THAT AFFECT THE RETINA

Some of the significant retinal disorders are the following: diabetic retinopathy, macular degeneration, retinal detachment, floaters, glaucoma, and ROP [1, 2]. Diabetic retinopathy is a condition caused by complications of diabetes mellitus. Diabetes mellitus is also the main cause of several other pathological conditions such as glaucoma and cataract. Diabetic retinopathy is also the prototype of a class of diseases, which can be described as chronic arteriolar capillaropathies. Among the conditions which may be included in this class are the retinopathies of hypertension, retinal vein occlusion, and possibly the degeneration of the macula [1, 3]. The malignant form of diabetic retinopathy is the chief cause of blindness due to diabetes.

Glaucoma is a term used to describe a number of conditions which are characterized by raised intraocular pressure, cupping of the ONH, and visual field defects. The pathology lies at the ONH, which becomes ischemic as a result of an imbalance between the intraocular pressure and the perfusion pressure in the capillaries of the retina [2].

ROP is a disorder of the developing blood vessels in the retina of very premature infants [4, 5, 6]. The most important characteristic of the ROP is abnormal dilation and tortuosity of posterior retinal vessels. The current treatment for advanced ROP includes laser photocoagulation, which can dramatically reduce the risk of permanent vision loss [7].

1.1.3   IMAGING OF THE RETINA

Early detection and treatment of various types of retinopathy are crucial to avoid preventable vision loss. Imaging of the retina began with human retinal photography [1]. Fluorescein angiography is an imaging method used in the diagnosis and management of retinal diseases [1]. However, fluorescein angiography requires the injection of fluorescein into the blood. With the recent advances in information and communications technology, improved techniques for digital photography of the retina have been developed. Digital fundus images of the retina are now an essential means to document and diagnose various eye diseases in clinics, and they are widely used to screen large populations for systemic diseases, such as diabetes, arteriosclerosis, and hypertension. Figure 1.1 shows a patient being imaged with a digital fundus camera.

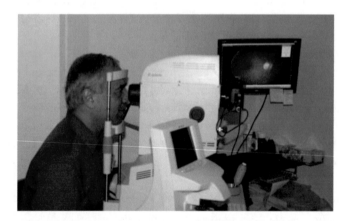

Figure 1.1: Imaging setup to obtain the fundus image of a patient.

The obvious advantages of digital images of the retina are the permanence of the record, the facilities for observer control, and the opportunity for precise comparison of the states of the retinal fundus at particular intervals of time. Additional advantages are the possibilities for precise measurement of the diameter of vessels and for observation of blood flow. Figure 1.2 is an example of a fundus image of a normal eye from the Digital Retinal Images for Vessel Extraction (DRIVE) dataset [8, 9]. The images in this dataset were acquired using a Canon CR5 non-mydriatic 3CCD camera with a 45 degree field of view (FOV). As an example, a basic package of a digital fundus imaging system contains a CR5-NM retinal camera, database software, and a data storage device.

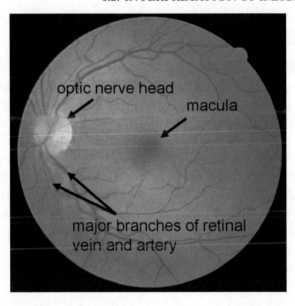

Figure 1.2: A macula-centered digital image of the retina from the DRIVE dataset (test image 01).

1.2 INTERPRETATION OF IMAGES OF THE RETINA

The color of the retinal fundus varies naturally according to the color of the light employed; however, the color of the normal fundus may be described as ranging from orange to vermillion [1]. The main features in a normal fundus image, including the ONH, macula, and major branches of the retinal artery and vein, are shown in Figure 1.2. The ONH is round or vertically oval in shape, measuring about 2 mm in diameter, and typically brighter than the surrounding area. The ONH is where retinal nerve fibers pass out of the globe of the eye to form the trunk of the optic nerve. Approximately 5 mm from the center of the ONH (equal to about 2.5 times the diameter of the ONH), can be seen the slightly oval-shaped, blood-vessel-free reddish spot, the fovea, which is at the center of the area known as the macula.

From the center of the ONH radiate the major branches of the retinal artery and vein, to the four quadrants of the retina. There is as much variation in the arterial pattern as in the venous pattern. In the macular region, all the vessels arch around, sending only small branches towards the avascular foveal area. The arteries appear with a brighter red color and are slightly narrower than the veins [1, 2]. Fundus images offer unique means for the examination of a part of the vascular tree and for the early diagnosis of diseases and conditions affecting the eye.

1.3 COMPUTER-AIDED DIAGNOSIS OF DISEASES AFFECTING THE EYE

Screening programs for retinopathy are in effect at many health-care centers around the world [10, 11]. Such programs require large numbers of retinal fundus images to be analyzed for the presence of diseases. Ophthalmologists examine retinal images, search for possible anomalies, and give the diagnostic results. Computer-aided diagnosis (CAD) via processing and analysis of retinal images by the application of image processing, computer vision, and pattern analysis techniques could provide a number of benefits. CAD can reduce the workload and provide objective decision-making tools to ophthalmologists. Quantitative analysis of the vascular architecture of the retina, as well as the changes in the shape, width, and tortuosity of the blood vessels could assist in the monitoring of the effects of diabetes, arteriosclerosis, hypertension, and premature birth on the visual system [10, 12]. The detection of abnormalities, such as aneurysms, exudates, and hemorrhages can assist in the diagnosis of retinopathy [10, 13, 14]. In Figure 1.3, an image of the retina from the Structured Analysis of the Retina (STARE) dataset [15, 16, 17] is shown, with exudates and clumping of retinal pigment epithelial (RPE) cells due to age-related macular degeneration. Four more examples of fundus images with pathology are shown in Figure 1.4: parts (a) and (b) in the figure show examples with exudates and hemorrhages; parts (c) and (d) of the same figure show two images with tortuous vessels.

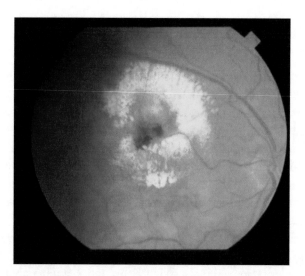

Figure 1.3: Image im0017 from the STARE dataset with exudates and clumping of RPE cells due to age-related macular degeneration.

A crucial preliminary step in computer-aided analysis of retinal images is the detection and localization of important anatomical structures, such as the ONH, the macula, the fovea, and the

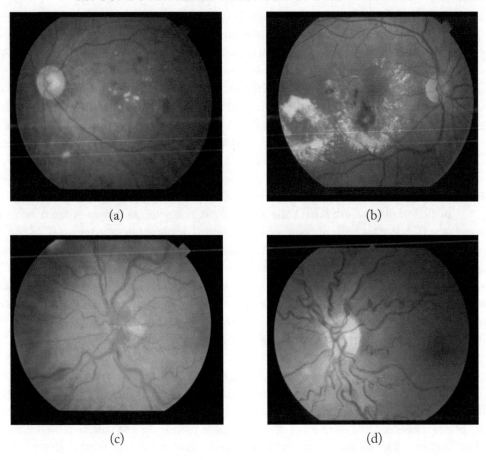

Figure 1.4: Images (a) im0009 and (b) im0049 from the STARE dataset with exudates and hemorrhages. Images (c) im0025 and (d) im0198 with tortuous vessels.

major vascular arcades; several researchers have proposed methods for these purposes [11, 13, 18, 19, 20]. The ONH and the major vascular arcades may be used to localize the macula and fovea [11, 13, 18]. The anatomical structures mentioned above may be used as landmarks to establish a coordinate system of the retinal fundus [11, 13, 18, 21]. Such a coordinate system may be used to determine the spatial relationship of lesions, edema, and hemorrhages with reference to the ONH and the macula [11]; it may also be used to exclude artifacts in some areas and pay more importance to potentially pathological features in other areas [13]. For example, when searching for microaneurysms and nonvascular lesions, the ONH area should be omitted due to the observation that the ONH area contains dot-like patterns, which could mimic the appearance of pathological features and confound the analysis [13]. Regardless, the position and average diameter of the ONH are used to grade new vessels and fibrous proliferation in the retina [21]. On the other hand, the presence of pathology in the macular region is associated with worse prognosis than elsewhere [13]; more attention could be paid by lowering the threshold of detection of pathological features in the macular region. Abundant presence of drusen (a characteristic of age-related macular degeneration or AMD) near the fovea has been found to be roughly correlated with the risk of wet AMD and the degree of vision loss [22]. One of the features useful in discriminating between drusen and hard exudates is that drusen are usually scattered diffusely or clustered near the center of the macula; on the other hand, hard exudates are usually located near prominent microaneurysms or at the edges of zones of retinal edema [21]. Criteria for the definition of clinically significant macular edema include retinal thickening with an extent of at least the average area of the ONH, with a part of it being within a distance equal to the average diameter of the ONH from the center of the macula [21].

In clinical diagnosis of AMD, the potentially blinding lesion known as choroidal neovascular membrane (CNVM) is typically observed as subfoveal, juxtafoveal, or extrafoveal, within the temporal vascular arcades. The CNVM lesion is usually circular in geometry and white in color. In a computer-aided procedure, the differentiation between the ONH and a potential CNVM lesion would be critical for accurate diagnosis, whereas both of these regions could be of similar shape and color, the ONH has converging vessels that dominate its landscape; hence, the prior identification of the ONH is important. Based on the same property noted above, one of the methods described in the present book for the detection of the ONH relies on the converging pattern of the vessels within the ONH.

Computer-aided analysis of retinal images has the potential to facilitate quantitative and objective analysis of retinal lesions and abnormalities. Different types of retinal pathology, such as ROP [7], diabetic retinopathy [10], and AMD [3] may be detected and analyzed via the application of appropriately designed techniques of digital image processing and pattern recognition. Accurate identification and localization of retinal features and lesions could contribute to improved diagnosis, treatment, and management of retinopathy.

1.4   SCOPE AND ORGANIZATION OF THE BOOK

The main aim of the work underlying the book is the development of digital image processing and pattern analysis techniques for the detection of the ONH in fundus images of the retina. The contents of the book are organized in seven chapters and a list of references.

Chapter 2 presents a brief discussion on computer methods for the detection of the main anatomical features of the retina in fundus images and the detection of abnormal features.

Chapter 3 presents an overview of image processing techniques that are essential in order to address the stated problem, including morphological filters, edge extraction, detection of oriented structures, and detection of geometrical patterns.

Chapter 4 gives details of the datasets used in the present work and the methods used for the evaluation of the results of detection.

Chapter 5 presents detailed descriptions of the procedures for and the results of the detection of the ONH using the Hough transform.

Chapter 6 provides detailed descriptions of the procedures for and the results of the detection of the ONH using Gabor filters and phase portraits.

Chapter 7 presents concluding remarks on computer-aided analysis of images of the retina.

CHAPTER 2

Computer-aided Analysis of Images of the Retina

2.1 DETECTION OF ANATOMICAL FEATURES OF THE RETINA

Computer-aided diagnosis of diseases affecting the eye, introduced in Section 1.3, requires the preliminary detection and delineation of the normal anatomical features of the retina, including the ONH, macula, and blood vessels [11, 18, 23, 24] (as shown in Figure 1.2). The locations of the main anatomical features can be taken as references or landmarks in the image for the detection of abnormal features. They may also be useful to mask out the normal anatomical features in the image to detect successfully the abnormal structures that could be present in an image of the retina; for example, an anatomical structure with a bright appearance located within the ONH could be ignored when detecting white lesions.

2.1.1 DETECTION OF THE OPTIC NERVE HEAD

The ONH appears toward the left-hand or right-hand side of a fundus image as an approximately circular area, roughly one-sixth the width of the image in diameter, brighter than the surrounding area, and as the convergent area of the network of blood vessels [1, 25]; see Figure 1.2. In an image of a healthy retina, all of the properties mentioned above (shape, color, size, and convergence) contribute to the identification of the ONH. A review of selected recent works on methods and algorithms to locate the ONH in images of the retina is provided in the following paragraphs.

Property-based methods: Based on the brightness and roundness of the ONH, Park et al. [25] presented a method using algorithms which include thresholding, detection of object roundness, and detection of circles. The successful detection rate obtained was 90.25% with the 40 images in the DRIVE dataset [8, 9]. Similar methods have been described by Barrett et al. [26], ter Haar [27], and Chrástek et al. [28, 29]. Sinthanayothin et al. [23] located the ONH by identifying the area with the highest variation in intensity using a window of size equal to that of the ONH. The images were preprocessed using an adaptive local contrast enhancement method applied to the intensity component. The method was tested with 112 images obtained from a diabetic screening service; a sensitivity of 99.1% was achieved.

Matched filter: In the work of Osareh et al. [30], a template image was created by averaging the ONH region of 25 color-normalized images. After locating the center of the ONH by using the template, gray-scale morphological filtering and active-contour modeling were used to locate the ONH region. An average accuracy of 90.32% in locating the boundary of the ONH was reported, with 75 images of the retina.

The algorithm proposed by Youssif et al. [31] is based on matching the expected directional pattern of the retinal blood vessels in the vicinity of the ONH. A direction map of the segmented retinal vessels was obtained by using a two-dimensional (2D) Gaussian matched filter. The minimum difference between the matched filter and the vessels' directions in the surrounding area of each ONH center candidate was found. The ONH center was detected correctly in 80 out of 81 images (98.77%) from a subset of the STARE dataset and all of the 40 images (100%) of the DRIVE dataset. A similar method has been implemented by ter Haar [27].

A template matching approach was implemented by Lalonde et al. [32]. The design relies on a Hausdorff-based template matching technique using edge maps, guided by pyramidal decomposition for large-scale object tracking. The proposed methods were tested with a dataset of 40 fundus images of variable visual quality and retinal pigmentation, as well as of normal and small pupils. An average error of 7% in positioning the center of the ONH was reported.

Geometrical model: The method proposed by Foracchia et al. [33] is based on a preliminary detection of the major retinal vessels. A geometrical parametric model, where two of the model parameters are the coordinates of the ONH center, was proposed to describe the general direction of retinal vessels at any given position. Model parameters were identified by means of a simulated annealing optimization technique. The estimated values provided the coordinates of the center of the ONH. An evaluation of the procedure was performed using a set of 81 images from the STARE dataset, containing both normal and pathological images. The position of the ONH was correctly identified in 79 out of the 81 images (97.53%).

Fractal-based method: Ying et al. [34] proposed an algorithm to differentiate the ONH from other bright regions such as hard exudates and artifacts, based on the fractal dimension related to the converging pattern of the blood vessels. The ONH was segmented by local histogram analysis. The scheme was tested with the DRIVE dataset and identified the ONH in 39 out of 40 images (97.5%).

Warping and random sample consensus: A method was proposed by Kim et al. [35] to analyze images obtained by retinal nerve fiber layer photography. In their method, the center of the ONH was selected as the brightest point and an imaginary circle was defined. Applying the random sample consensus technique, the imaginary circle was first warped into a rectangle and then inversely warped into a circle to find the boundary of the ONH. The images used to test the method included 43 normal images and 30 images with glaucomatous changes. The reported performance of the algorithm was 91% sensitivity and 78% positive predictability.

Convergence of blood vessels: Hoover and Goldbaum [16] used fuzzy convergence to detect the origin of the blood-vessel network, which can be considered as the center of the ONH in a fundus image. The method was tested using 30 images of normal retinas and 51 images of retinas with pathology from the STARE dataset, containing such diverse symptoms as tortuous vessels, choroidal neovascularization, and hemorrhages that obscure the ONH. The rate of successful detection achieved was 89%. Fleming et al. [13] used the elliptical form of the major retinal blood vessels to obtain an approximate region of the ONH, which was then refined based on the circular edge of the ONH. The methods were tested on 1056 sequential images from a retinal screening program. In 98.4% of the cases tested, the error in the ONH location was less than 50% of the diameter of the ONH.

Tensor voting and adaptive mean-shift: The method proposed by Park et al. [36] was based on tensor voting to analyze vessel structures; the position of the ONH was identified by mean-shift-based mode detection. Park et al. used three preprocessing stages through illumination equalization to enhance local contrast and extract vessel patterns by tensor voting in the equalized images. The position of the ONH was identified by mode detection based on the mean-shift procedure. The method was evaluated with 90 images from the STARE dataset, and achieved 100% success rate on 40 images of normal retinas and 84% on 50 images of retinas with pathology.

Genetic algorithms: Carmona et al. [37] proposed a method to obtain an ellipse approximating the ONH using a genetic algorithm. The parameters characterizing the shape of the ellipse obtained were also provided by the algorithm. Initially, a set of hypothesis points were obtained that exhibited geometric properties and intensity levels similar to the ONH contour pixels. Next, a genetic algorithm was used to find an ellipse containing the maximum number of hypothesis points in an offset of its perimeter, considering some constraints. The method is designed to locate and segment the ONH in images of the retina without any intervention by the user. The method was tested with 110 images of the retina. The results were compared with a gold standard, generated from averaging the contours traced by different experts; the results for 96% of the images had less than five pixels of discrepancy. Hussain [38] proposed a method combining a genetic algorithm and active contour models; no quantitive result was reported.

Further comparative analysis of the methods reviewed above with the methods described in the book is provided in Section 5.4 and Section 6.4.

2.1.2 DETECTION OF THE MACULA

The macular region is generally darker than the surrounding retinal tissue, due to the higher density of carotenoid pigments in the retina and pigment granules in the retinal pigment epithelial layer beneath the retina. However, the macula exhibits nonspecific structure and varies greatly across individuals due to variations in the levels of pigment associated with factors such as ethnicity, age, diet, and disease states [1, 10]. Ancillary imaging techniques have been used to facilitate the detection of the macula, for example, Argonblue imaging to highlight pigmentation [39]. Techniques have been developed to locate the darker pigmented regions associated with the macula [11].

Many methods to identify the macula are based on the preliminary detection of the ONH and the vasculature [11, 18, 23], assuming that the macula is located at approximately 2.5 times the diameter of the ONH from the center of the ONH [1]. Sinthanayothin et al. [23] proposed a template matching approach to detect the macula. Li and Chutatape [11] presented a model-based approach, using an active shape model to extract the main course of the vasculature based on the location of the ONH. The locations of the vasculature and the ONH were used to identify the macula. Tobin et al. [18] applied a parabolic model to detect the curvature of the vasculature and the ONH. The fovea was located at a fixed distance from the ONH along the horizontal raphe of the retina.

2.1.3   DETECTION OF BLOOD VESSELS

The detection of blood vessels is an important step in applications of image processing in ophthalmology [8, 10, 11, 33, 40, 41, 42, 43]. Retinal vascular segmentation techniques utilise the contrast existing between the retinal blood vessels and the surrounding background. The cross-sectional gray-level profile of a typical vessel conforms to a Gaussian shape. The vasculature is piecewise linear and may be represented by a series of connected line segments. Five main techniques used to segment the vasculature in retinal images are matched filters [16, 40, 42], Gabor filters [12, 43], vessel tracking [44], neural networks [23], and morphological processing [45, 46].

There are many research works published on the detection of retinal vessels within the ONH and the estimation of specific vessel measurements, including the changes in shape, width, and tortuosity, that reveal information related to the stage and effects of conditions, such as glaucoma, diabetes, hypertension, and ROP [8, 39, 40, 43, 46, 47, 48, 49].

2.2   DETECTION OF ABNORMAL FEATURES

Abnormal features in images of the retina include aneurysms, exudates, and hemorrhages associated with diabetic retinopathy, glaucoma, ROP, hypertension, and other types of retinopathy [1, 10] (as shown in Figure 1.4).

Research works on automated detection of diabetic retinopathy in digital images of the retina have concentrated largely on the automated detection of aneurysms, exudates, and hemorrhages [13, 50, 51]. Methods designed to detect abnormal features focus on distinguishing them from a number of different distractors, such as small vessels, choroidal vessels, and reflection artifacts.

Exudates are one of the commonly occurring lesions [1]. Three strategies have been employed to detect exudates: thresholding [52], edge detection [11], and classification [53]. The thresholding method is a straightforward approach to detect exudates; however, the selection of the threshold is difficult due to the uneven intensity of the exudates, and the low contrast between exudates and the retinal background. Methods based on edge detection or classification can overcome this difficulty and assist in CAD.

2.3   LONGITUDINAL ANALYSIS OF IMAGES

Methods for longitudinal analysis of images of the retina are designed to facilitate follow-up of patients with retinal pathology and to perform assessment of the effect of treatment [7, 41]. Narasimha-Iyer et al. [41] described algorithms, including Bayesian classification and illumination correction, to detect and classify changes in time series of color fundus images of the retina. Ells et al. [7] evaluated the use of remote reading of digital retinal photographs in the diagnosis of severe ROP in a program for longitudinal screening for ROP.

2.4   REMARKS

A general review was presented in this chapter on computer methods for the detection of both normal and abnormal features in fundus images of the retina. Specific details of the methods developed in the book for the detection of the ONH are described in the subsequent chapters. The performance of the methods described is evaluated in comparison with some of the works reviewed in this chapter.

CHAPTER 3

Detection of Geometrical Patterns

The images in the DRIVE and STARE datasets are provided in the $[R, G, B]$ format, where R, G, and B are the red, green, and blue components, respectively, of the color image. In the present work, after normalizing each component of the original color image (dividing by 255), the result was converted to the luminance component Y, computed as $Y = 0.299R + 0.587G + 0.114B$. The effective region of the image was thresholded using the normalized threshold of 0.1 for the DRIVE images and 0.15 for the STARE images; the thresholds were determined by experimentation with several images from each dataset. The artifacts present in the thresholded results on the edges were removed by applying morphological opening and erosion filters. A mask was generated with the obtained effective region.

3.1   FILTERS FOR PREPROCESSING IMAGES

Median and mean filters: Median and mean filters are space-domain local-statistics-based filters, commonly used in digital image processing to remove noise and smooth images [54, 55]. In the case of the mean filter, the mean of the pixels in a specified neighborhood of the pixel being processed is used as the output. If the median value of the set of pixels in the chosen neighborhood is taken, the filter is called the median filter. Different spatial configuration of the neighborhood used will result in different effects of the mean and median filters [55]. The median filter provides better noise removal than the mean filter without blurring, especially in cases where outliers exist in the image, and particularly in the case of salt-and-pepper noise. In the preprocessing step of the present work, the median filter was used to remove outliers, which were observed as bright isolated points in images of the retina.

Morphological filters: Morphological filters [54, 56] include the basic operations of erosion and dilation, and modifications and combinations of these operations. Fundamentally, all morphological filters are neighborhood-based operators with a structuring element for the neighborhood. Morphological filters applied to binary images, discussed in the following paragraphs, are simpler and usually involve counting pixels rather than weighted multiplication and addition as in the case of filters applied to gray-scale images; this is because the values of the pixels in binary images are restricted to 0 or 1.

Erosion removes pixels from an image, or, equivalently, turns pixels that were originally 1 to 0. The simplest kind of erosion, sometimes referred to as classical erosion, is to remove any pixel touching another pixel that is part of the background. This removes a layer of pixels from the periphery of all features and regions, which will cause some shrinking. Erosion will remove extraneous pixels altogether because such artifacts are normally only a single pixel wide. In Figure 3.1, a binary test image is shown in part (a), and the result of applying erosion with a disc-shaped structuring element of diameter 5 pixels is shown in part (b). We can observe that the two small isolated circular objects have been removed because they are smaller than the structuring element. The holes within the larger object have become bigger. Furthermore, the narrow link in the large object has been disconnected. A ribbon of width 5 pixels has been removed from the periphery of all features and regions.

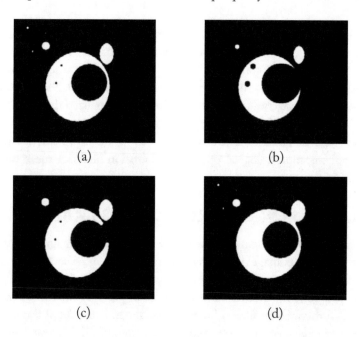

(a)   (b)

(c)   (d)

Figure 3.1: (a) A binary test image. The size of the image is 307 × 368 pixels. (b) Result after applying erosion with a disc-shaped structuring element of diameter 5 pixels. (c) Result after applying an opening filter, which removes objects or regions in the image under diameter of 5 pixels. (d) Result after applying a closing filter, which fills in holes in the image under the diameter of 5 pixels.

Instead of removing pixels from features, the complementary operation known as dilation can be used to add pixels. This operation adds a layer of pixels to the periphery of all features and regions; it also fills in small holes within features.

The combination of an erosion followed by a dilation is called an opening operation, referring to the ability of this combination to open up spaces between just-touching features. It is one of the most commonly used sequences of morphological operations to remove noise from binary images. If the combination is performed in the opposite order, that is, a dilation followed by an erosion, the operation is called a closing. The result is to fill in places where isolated pixels are off, missing pixels within features, or narrow gaps between portions of a feature. In Figure 3.1 (c), the result of applying an opening operator with a disc-shaped structuring element of diameter 5 pixels is shown. The operator disconnects the narrow part of the largest object. Also, small isolated objects are removed. As shown in Figure 3.1 (d), applying a closing operator results in small holes and the narrow gap being filled.

In the present work, the artifacts present at the edges of the masks generated for images from the DRIVE and STARE datasets were removed by applying morphological erosion with a disc-shaped structuring element of diameter 5 pixels. In Figure 3.2 (a), the STARE image im0050 is shown; the corresponding Y component is shown in part (b). Part (c) of the same figure shows the result thresholded at 0.15 of the normalized intensity, and part (d) shows the result of applying morphological erosion with a disc-shaped structuring element of diameter 5 pixels and morphological opening to the thresholded result in part (c).

Preprocessing of the fundus images: In order to avoid edge artifacts in subsequent steps, each image was extended beyond the limits of its effective region [12, 43]. First, a four-pixel neighborhood was used to identify the pixels at the outer edge of the effective region. For each of the pixels identified, the mean gray level was computed over all pixels in a 21×21 neighborhood that were also within the effective region and assigned to the corresponding pixel location. The effective region was merged with the outer edge pixels, forming an extended effective region. The procedure was repeated 50 times, extending the image by a ribbon of width 50 pixels.

3.2   DETECTION OF EDGES

Edge detection plays an important role in a number of digital image processing applications such as object recognition. The edges in an image provide useful structural information about object boundaries because edges are caused by changes in some physical properties of the surfaces being imaged, such as illumination, geometry, and reflectance. An edge is, therefore, defined as a local change or discontinuity in image luminance. In edge detection methods, the discontinuities in the image gray level are enhanced by neighborhood operators. Prewitt and Sobel operators [54, 55, 57] use masks having the size of 3×3 pixels.

3.2.1   PREWITT AND SOBEL OPERATORS

The Prewitt operators for the horizontal and vertical derivatives, $G_x(x, y)$ and $G_y(x, y)$, respectively, are defined as follows:

$$
\begin{bmatrix}
-1 & 0 & 1 \\
-1 & 0 & 1 \\
-1 & 0 & 1
\end{bmatrix}, \tag{3.1}
$$

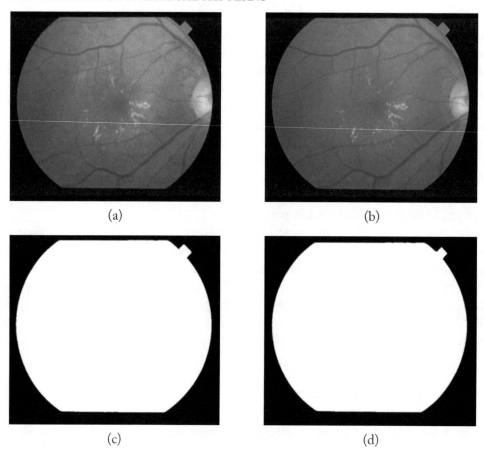

Figure 3.2: (a) Image im0050 of the STARE dataset. Image size: 700×605 pixels. (b) Y component of the original color image. (c) Binary image obtained from (b) thresholded at 0.15 of the normalized intensity. (d) The mask generated by applying morphological erosion with a disc-shaped structuring element of diameter 5 pixels and morphological opening to remove objects under $10,000$ pixels in the result in part (c).

$$\begin{bmatrix} -1 & -1 & -1 \\ 0 & 0 & 0 \\ 1 & 1 & 1 \end{bmatrix}.$$

(3.2)

The Sobel operators are similar to the Prewitt operators, but they include larger weights for the pixels in the row and column of the pixel being processed. The Sobel operators for the horizontal and vertical derivatives, $G_x(x, y)$ and $G_y(x, y)$, respectively, are defined as follows:

$$\begin{bmatrix} -1 & 0 & 1 \\ -2 & 0 & 2 \\ -1 & 0 & 1 \end{bmatrix}, \tag{3.3}$$

$$\begin{bmatrix} -1 & -2 & -1 \\ 0 & 0 & 0 \\ 1 & 2 & 1 \end{bmatrix}. \tag{3.4}$$

The horizontal and vertical components of the gradient, $G_x(x, y)$ and $G_y(x, y)$, respectively, are obtained by convolving the preprocessed image with the corresponding Sobel or Prewitt operators. A commonly used definition of the combined gradient magnitude is

$$G(x, y) = [G_x^2(x, y) + G_y^2(x, y)]^{\frac{1}{2}}. \tag{3.5}$$

By thresholding the gradient magnitude, a binary edge map can be obtained. Different values of the threshold will result in different numbers of nonzero pixels in the edge map. An optimal threshold may need to be determined according to the intensities of the edges to be detected in the image. The preprocessed version of image 01 from the DRIVE dataset is shown in Figure 3.3 (a). The horizontal and vertical components of the Sobel gradient are shown in parts (b) and (c) in Figure 3.3. The operators enhance the horizontal and vertical edges in the image and give high values (positive or negative) around the outer edges of the blood vessels. The combined gradient magnitude is shown in part (d) and the binary image after thresholding at 0.18 of the normalized intensity is shown in part (e). The edges of the blood vessels and the rim of the ONH are clearly seen in the edge map shown in part (e).

3.2.2 THE CANNY METHOD

Canny [59] proposed an approach for edge detection based upon three criteria for good edge detection, including multidirectional derivatives, multiscale analysis, and optimization procedures [55]. Canny's method is guided by the following requirements:

- The edge detector should have a high probability of finding all edges and low probability of mistaking non-edge objects as edges, represented in the form of a signal-to-noise ratio.

- The edges found by the algorithm should also be as close to the true edges as possible, represented by the root mean-squared distance of the detected edges from the true edges.

- The algorithm should have only one edge pixel when an edge is present, represented by the distance between the adjacent maxima in the output.

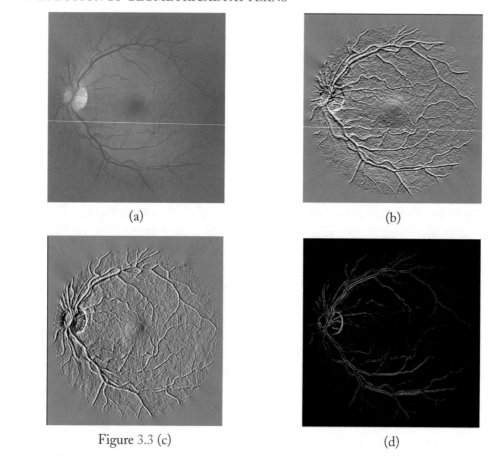

(a)   (b)

Figure 3.3 (c)   (d)

Figure 3.3: *Continues.*

The Canny method implements edge detectors based upon kernel functions formed by using the first derivative of Gaussian functions. The output of the Canny method can be thresholded to obtain an edge map. The MATLAB [58] version of the Canny method was used to obtain a binary edge map. The procedure requires the specification of three parameters: the standard deviation of the Gaussian and two thresholds. In the present work, the default values of the parameters provided by the MATLAB functions were used (standard deviation = 1 and the thresholds determined automatically). The resulting edge image is shown in Figure 3.3 (f) for image 01 of the DRIVE dataset. The threshold for the edge map was set to be 0.18 of the normalized intensity for the illustration. Comparing the output of the Sobel operators and the Canny method in Figures 3.3 (e) and (f), we can observe that the Canny method gives connected edges whereas the Sobel operators give disconnected edges. The edge map resulting from the Canny method is much cleaner than that from the Sobel operators,

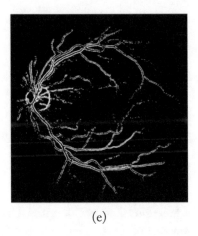
(e)

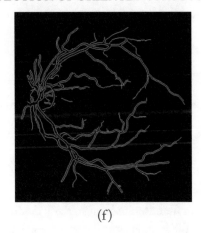
(f)

Figure 3.3: *Continued.* (a) Preprocessed luminance component of image 01 of the DRIVE dataset. The original color image is shown in Figure 1.2. (b) Horizontal component of the Sobel gradient of the image; the range [-0.1, 0.1] has been mapped to [0, 255]. (c) Vertical component of the gradient of the Sobel image; the range [-0.1, 0.1] has been mapped to [0, 255]. (d) The combined Sobel gradient magnitude image; the range [0, 0.1] has been mapped to [0, 255]. (e) The binary image obtained from (d); threshold = 0.18 of the normalized intensity. (f) The binary image obtained using the Canny method; threshold = 0.18 of the normalized intensity. The MATLAB [58] version of the Canny method was used.

without large numbers of isolated points resulting from the small vessels present in the image. However, the Canny method is computationally more complex than the Sobel operators because of its optimization procedures.

3.3   DETECTION OF ORIENTED STRUCTURES

3.3.1   THE MATCHED FILTER

Matched filters [8, 15, 40, 60] may be used efficiently to extract certain objects of interest in a given image, with the spatial properties of the object to be detected being known in advance. Matched filters have been applied for the detection of blood vessels in fundus images of the retina [8, 15, 40, 60], based upon a model assuming cross-sections of blood vessels to have a shape that is a Gaussian function.

The design of matched filters is based on the following properties related to the appearance of blood vessels in fundus images:

- Vessels can be approximated by pairs of antiparallel segments.

- Vessels have lower reflectance than other retinal surfaces, and hence appear to be relatively darker than the background.

- The size of a vessel decreases when moving away from the ONH.

- The width of a retinal vessel lies within a certain known range.

- The intensity profiles vary by a small amount from vessel to vessel.

- The intensity profile has a Gaussian shape.

The output of a matched filter is a gray-scale image which requires a thresholding procedure to detect blood vessels.

3.3.2   THE GABOR FILTER

Gabor functions are sinusoidally modulated Gaussian functions that provide optimal localization in both the frequency and space domains; a significant amount of research has been conducted on the use of Gabor functions or filters for segmentation, analysis, and discrimination of various types of texture [55, 61, 62, 63].

The real Gabor filter kernel oriented at the angle $\theta = -\pi/2$ may be formulated as [61, 62]

$$g(x, y) = \frac{1}{2\pi \sigma_x \sigma_y} \exp\left[-\frac{1}{2}\left(\frac{x^2}{\sigma_x^2} + \frac{y^2}{\sigma_y^2} \right) \right] \cos(2\pi f_o x), \tag{3.6}$$

where σ_x and σ_y are the standard deviation values in the x and y directions, and f_o is the frequency of the modulating sinusoid. Kernels at other angles are obtained by rotating the basic kernel over the range $[-\pi/2, \pi/2]$ by using the co-ordinate transformation

$$\begin{bmatrix} x' \\ y' \end{bmatrix} = \begin{bmatrix} \cos\alpha & \sin\alpha \\ -\sin\alpha & \cos\alpha \end{bmatrix} \begin{bmatrix} x \\ y \end{bmatrix}, \tag{3.7}$$

where (x', y') is the set of coordinates rotated by the angle α. In the present work, a set of 180 kernels was used, with angles spaced evenly over the range $[-\pi/2, \pi/2]$.

The Gabor filters may be used as line detectors [62]. The parameters in Equation 3.6, namely $\sigma_x, \sigma_y,$ and f_o, need to be specified by taking into account the size of the lines or curvilinear structures to be detected. Let τ be the thickness of the line detector. This parameter is related to σ_x and f_o as follows [12, 61, 62]:

- The amplitude of the exponential (Gaussian) term in Equation 3.6 is reduced to one half of its maximum at $x = \tau/2$ and $y = 0$; therefore, $\sigma_x = \tau/(2\sqrt{2\ln 2}) = \tau/2.35$.

- The cosine term has a period of τ; hence, $f_o = 1/\tau$.

- The value of σ_y could be defined as $\sigma_y = l\,\sigma_x$, where l determines the elongation of the Gabor filter in the orientation direction, with respect to its thickness. In the present work, $l = 2.9$ was used.

The value of τ could be varied to prepare a bank of filters at different scales for multiresolution filtering and analysis [12]; however, in the present work, a single scale is used. The effects of the different design parameters are shown in Figure 3.4.

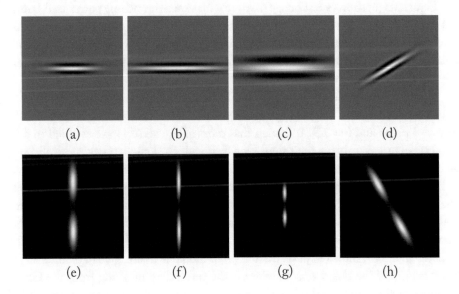

Figure 3.4: Effects of the different parameters of the Gabor filter. (a) Example of the impulse response of a Gabor filter. (b) The parameter l is increased: the Gabor filter is elongated in the x direction. (c) The parameter τ is increased: the Gabor filter is enlarged in the x and y directions. (d) The angle of the Gabor filter is modified. Figures (e) – (h) correspond to the magnitude of the Fourier transforms of the Gabor filters in (a) – (d), respectively. The (0, 0) frequency component is at the center of the spectra displayed. Reproduced with permission from F.J. Ayres and R.M. Rangayyan, "Characterization of architectural distortion in mammograms via analysis of oriented texture", *IEEE Engineering in Medicine and Biology Magazine*, 24(1): 59-67, January 2005. © IEEE.

The Gabor filter designed as above can detect piecewise-linear features of positive contrast, for example, linear elements that are brighter than their immediate background. In the present work, because the blood vessels are darker than the background, the Gabor filter was applied to the inverted version of the preprocessed image.

Blood vessels in the retina vary in thickness in the range $50 - 200 \ \mu$m, with a median of $60 \ \mu$m [4, 10]. Taking into account the size and the spatial resolution of the images in the DRIVE and STARE datasets, the parameters for the Gabor filters were specified as $\tau = 8$ pixels and $l = 2.9$ in the present work. For each image, a magnitude response image was composed by selecting the maximum response over all of the Gabor filters for each pixel. An angle image was prepared by using the angle of the filter with the largest magnitude response for each pixel. Each angle image was filtered with a Gaussian filter having a standard deviation of 6 pixels and downsampled by a factor of 4 for subsequent analysis; the result is referred to as the orientation field. The parameters listed above were determined by experimentation with a number of images from the DRIVE and STARE datasets, and by taking into consideration the average thickness of the blood vessels in the retina.

Figure 3.5 (a) shows a test image of a tree with several branching patterns. A bank of 180 Gabor filters with $\tau = 8$ pixels and $l = 2.9$ were used. Part (b) of the same figure shows the magnitude response of the Gabor filters, with high responses at major branches. Part (c) of the same figure shows the orientation field derived from the Gabor filters. The needles indicating the local orientation field follow the directions of the tree branches.

Figure 3.6 (a) shows an image (im0255) from the STARE dataset. The result in part (b) of the same figure shows the magnitude response of the Gabor filters, using a bank of 180 Gabor filters with $\tau = 8$ pixels and $l = 2.9$. It is seen that most of the vessels have resulted in a high response. Part (c) of Figure 3.6 shows the orientation field, which demonstrates successful mapping of the local orientation of the vascular structures in the image. (Needles indicating the local orientation have been drawn for every fifth pixel in the row and column directions.) The magnitude response of the Gabor filters, after being thresholded to a binary image, can be used to detect blood vessels in the image. In Figure 3.7 (d), the vessel map, resulting from thresholding the Gabor magnitude response, is shown for image 01 from DRIVE database. The edge maps resulting from the Sobel and Canny methods, shown in parts (b) and (c) of the same figure, give high-intensity values around the outer edges of the blood vessels, whereas the vessel map resulting from the Gabor filters gives high values at the center of the blood vessels. Figure 3.6 (a) shows an image (im0255) from the STARE dataset. The result in part (b) of the same figure shows the magnitude response of the Gabor filters, using a bank of 180 Gabor filters with $\tau = 8$ pixels and $l = 2.9$. It is seen that most of the vessels have resulted in a high response. Part (c) of Figure 3.6 shows the orientation field, which demonstrates successful mapping of the local orientation of the vascular structures in the image. (Needles indicating the local orientation have been drawn for every fifth pixel in the row and column directions.) The magnitude response of the Gabor filters, after being thresholded to a binary image, can be used to detect blood vessels in the image. In Figure 3.7 (d), the vessel map resulting from thresholding the Gabor magnitude response is shown for image 01 from DRIVE database. The edge maps resulting from the Sobel and Canny methods, shown in parts (b) and (c) of the same figure, give high-intensity values around the outer edges of the blood vessels, whereas the edge map resulting from the Gabor filters gives high values at the center of the blood vessels.

3.4 DETECTION OF GEOMETRICAL PATTERNS

3.4.1 THE HOUGH TRANSFORM

The Hough transform is a useful tool in digital image analysis for the recognition of parameterized geometrical shapes [54, 55, 64, 65]. The strength of the Hough transform lies in the ability to identify geometrical shapes existing in a given image even when the edges of the shapes are sparse or disconnected. The basic idea of the Hough transform is to find curves that can be parameterized, for example straight lines, parabolas, and circles, in a parameter space, which is also known as the

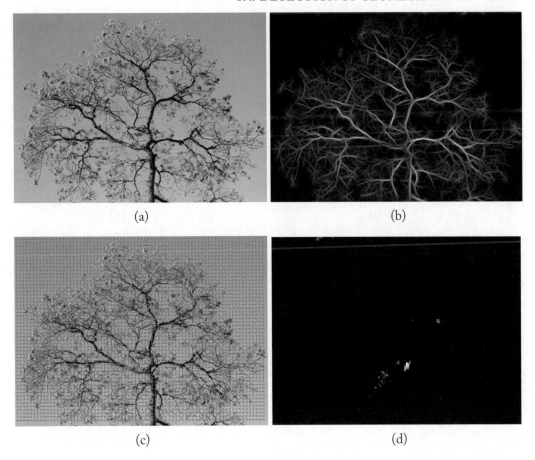

Figure 3.5: (a) Y component of a color image of a tree. Image size: 435×600 pixels. (b) Magnitude response of a bank of 180 Gabor filters. (c) Orientation field. Needles indicating the local orientation have been drawn for every sixth pixel in the row and column directions. (d) Node map. The intensity values correspond to the number of votes accumulated at each location. The intensity has been enhanced by using the log operation and the result has been mapped to [0, 255].

Hough space. The Hough transform can be used with two parameters to find straight lines, centers of circles with a fixed radius, or parabolas defined as $y = ax^2 + bx + c$ with constant c. The Hough transform can also be used in higher dimensions in the case of shapes with more parameters.

Detection of straight lines: Let us consider the problem of detection of straight lines in a binary image. Assume that the pixels that belong to a line have the value 1, and all other pixels have the value 0; assume also that the line is one-pixel thick. A straight line can be parameterized in the form:

$$\rho = x \cos \theta + y \sin \theta, \tag{3.8}$$

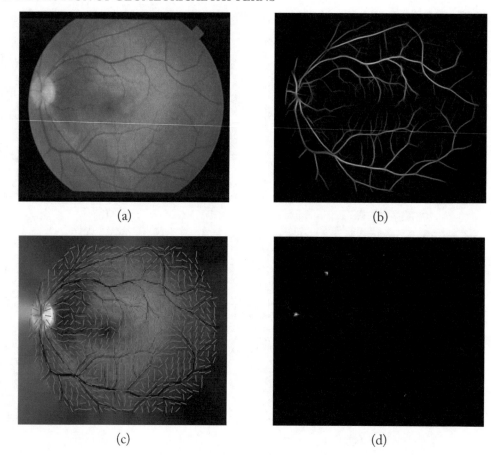

Figure 3.6: *Continues.*

where ρ is the perpendicular distance of the line from the origin and θ is the angle of the normal.

If the normal parameters of a line are (ρ_0, θ_0), all pixels along the line satisfy the relationship

$$\rho_0 = x(n) \cos \theta_0 + y(n) \sin \theta_0. \tag{3.9}$$

For a given pixel $[x(n), y(n)]$, this represents a sinusoidal curve in the (ρ, θ) parameter space; it follows that the curves for all the pixels of the specified line intersect at the point (ρ_0, θ_0). The associated peak value in the Hough space could be used to detect the line.

Figure 3.8 (a) shows an image with two straight lines represented by the parameters $(\rho, \theta) = (50, 30°)$ and $(\rho, \theta) = (-10, 60°)$. The limits of the x and y axes are ± 50, with the origin at the center of the image. Part (b) of the same figure shows the Hough space of the image. The range of θ is $[0, 180°]$; the range of ρ is $[-75, 75]$. The parameter space demonstrates the expected sinusoidal patterns, as well as two peaks at the locations corresponding to the parameters of the two lines present

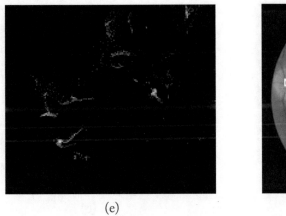
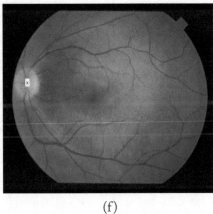

(e)   (f)

Figure 3.6: *Continued.* (a) STARE image im0255. Image size: 700 × 605 pixels. (b) Magnitude response of a bank of 180 Gabor filters. (c) Filtered and downsampled orientation field. Needles indicating the local orientation have been drawn for every fifth pixel in the row and column directions. (d) Node map. The intensity values correspond to the number of votes accumulated at each location. The intensity has been enhanced by using the log operation, and the result has been mapped to [0, 255]. (e) Saddle map. (f) The peak detected in the node map, located at (76, 252), superimposed on the original image. Reproduced with permission from R.M. Rangayyan, X. Zhu, F.J. Ayres, and A.L. Ells, "Detection of the optic nerve head in fundus images of the retina with Gabor filters and phase portrait analysis", *Journal of Digital Imaging*, 23(4): 438-453, August 2010. © Springer.

in the image. The intensity of the point at the intersection of the sinusoidal curves corresponding to the line with the parameters $(\rho, \theta) = (50, 30°)$ is less than that of the other line, reflecting its shorter length.

Detection of circles: The points lying on the circle

$$[x(n) - a]^2 + [y(n) - b]^2 = c^2 \tag{3.10}$$

are represented by a single point in the three-dimensional (3D) Hough space (a, b, c) with an accumulator of the form $A(a, b, c)$. Here, (a, b) is the center, and c is the radius of the circle. The procedure to detect circles involves the following steps:

1. Obtain a binary edge map of the image.

2. Set values for a and b.

3. For each nonzero pixel in the edge map, solve for the value of c that satisfies Equation 3.10.

4. Increment the accumulator that corresponds to (a, b, c).

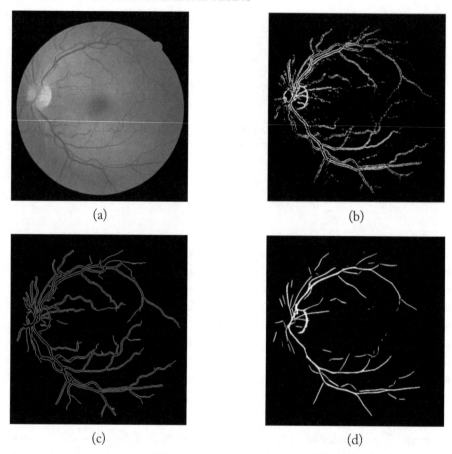

(a)   (b)

(c)   (d)

Figure 3.7: (a) DRIVE image 01. Image size: 584×565 pixels. (b) Edge map obtained using the Sobel operators thresholded at 0.18 of the normalized intensity. (c) Edge map obtained using the Canny method thresholded at 0.18 of the normalized intensity. (d) Vessel map obtained by thresholding the Gabor magnitude response at 0.009 of the normalized intensity.

5. Update values for a and b within the range of interest and go back to Step 3.

A test image is shown in Figure 3.9 (a) with a circle centered at (100, 150) and a radius of 50 pixels. The binary edge map for use as input to the Hough transform was obtained by using the Sobel operators, and is shown in part (b) of the same figure. The Hough space images (planes) with $c = 20$ and $c = 50$ pixels are shown in parts (c) and (d) of the same figure, respectively. A clear peak is seen in the Hough space for $c = 50$ pixels at the center of the circle. Another test image with two circles, located at (100, 150) and (150, 50), with radii of 50 and 20 pixels, respectively, is shown in Figure 3.10 (a). The input binary edge map to the Hough transform was obtained by using

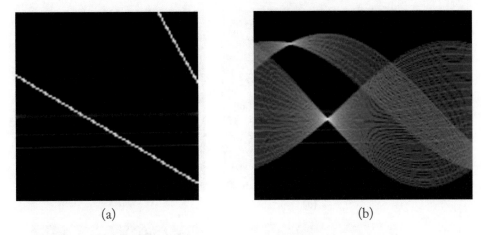

(a)   (b)

Figure 3.8: (a) Image with two straight lines with $(\rho, \theta) = (50, 30°)$ and $(\rho, \theta) = (-10, 60°)$. The limits of the x and y axes are ±50, with the origin at the center of the image. (b) Hough transform parameter space for the image. The display intensity is $\log(1 + accumulator\ cell\ value)$. The horizontal axis represents $\theta = [0, 180°]$; the vertical axis represents $\rho = [-75, 75]$.

the Sobel operators, and is shown in part (b) of the same figure. The Hough space images (planes) with $c = 20$ and $c = 50$ are shown in parts (c) and (d) of the same figure. We can observe a peak in Figure 3.10 (d) located at $(100, 150)$, which corresponds to the center of the circle with the radius of 50 pixels. In Figure 3.10 (c), the peak located at $(150, 50)$ corresponds to the center of the circle with the radius of 20 pixels. The intensity of the peak in Figure 3.10 (c) is less than that in part (d) of the same figure, in proportion to the number of points on the boundary of the corresponding circle.

The Hough space images (planes) for the testing image 01 in the DRIVE dataset with $c = 31$, $c = 37$, and $c = 47$ pixels are shown in Figure 3.11 (c), (d), and (e), respectively. The input binary edge map to the Hough transform was obtained by using the Sobel operators and is shown in part (b) of the same figure. Each local maximum in each plane of the Hough space corresponds to a possible circle with the corresponding radius and center in the original image. Several distinct peaks are visible in the planes shown. The global maximum of the Hough space is in the plane with $c = 37$ pixels, and the location of the peak gives the coordinates of the center of the circle as $(89, 260)$; this circle is shown overlaid on the original image in Figure 3.11 (f).

3.4.2   PHASE PORTRAITS

Phase portraits [66, 67] can be used for the analysis of images displaying oriented texture. The appearance of the phase portrait diagram can be categorized as node, saddle, or spiral [68]. In the present work, only the node and saddle phase portraits are used.

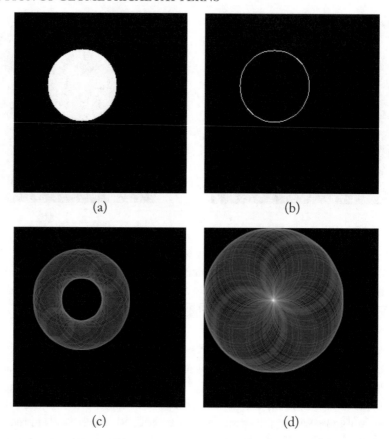

Figure 3.9: (a) Test image (250 × 250 pixels) with a circle with the center at (100, 150) and radius of 50 pixels. (b) Binary edge image of (a). (c) Hough space with $c = 20$ pixels. (d) Hough space with $c = 50$ pixels.

Consider the system of differential equations specified as

$$\begin{bmatrix} \dot{x}(t) \\ \dot{y}(t) \end{bmatrix} = \mathbf{A} \begin{bmatrix} x(t) \\ y(t) \end{bmatrix} + \mathbf{b}, \tag{3.11}$$

$$\mathbf{A} = \begin{bmatrix} a & b \\ b & c \end{bmatrix}, \quad \mathbf{b} = \begin{bmatrix} d \\ e \end{bmatrix}, \tag{3.12}$$

where $\dot{x}(t)$ and $\dot{y}(t)$ are the first-order derivatives of $x(t)$ and $y(t)$ with respect to time t, which represent the components of the velocity of a particle. $\mathbf{A}$ is a 2×2 matrix and $\mathbf{b}$ is a 2×1 column matrix (a vector). The functions $x(t)$ and $y(t)$ can be associated with the x and y coordinates of the

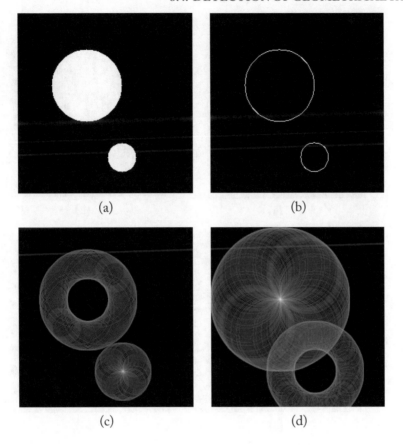

(a)   (b)

(c)   (d)

Figure 3.10: (a) Test image (250×250 pixels) with two circles: one with the center of $(100, 150)$ and radius of 50 pixels and the other with the center at $(150, 50)$ and radius of 20 pixels. (b) Binary edge image of (a). (c) Hough space with $c = 20$ pixels. (d) Hough space with $c = 50$ pixels.

rectangular coordinate plane of an image $f(x, y)$ [62, 66]. The orientation field generated by the phase portrait model is defined as

$$\phi(x, y | \mathbf{A}, \mathbf{b}) = \arctan\left(\frac{\dot{y}(t)}{\dot{x}(t)}\right), \tag{3.13}$$

which is the angle of the velocity vector $[\dot{x}(t), \dot{y}(t)]$ with the x axis at $(x, y) = [x(t), y(t)]$. The fixed point of the phase portrait is the point where $\dot{x} = \dot{y} = 0$, and denotes the center of the phase portrait pattern being observed. Let $\mathbf{x}_0 = [x_0, y_0]^T$ be the coordinates of the fixed point. From Equation 3.11, we have

$$\mathbf{x}_0 = \begin{bmatrix} x_0 \\ y_0 \end{bmatrix} = -\mathbf{A}^{-1}\mathbf{b}. \tag{3.14}$$

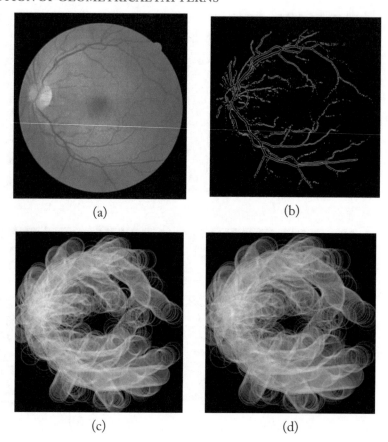

(a)

(b)

(c)

(d)

Figure 3.11: *Continues.*

According to the eigenvalues of **A** (the *characteristic matrix*), as shown in Table 3.1, the phase portrait is classified as a node, saddle, or spiral [68]. The formation of each phase portrait type is explained as follows [67, 68].

Node: The components $x(t)$ and $y(t)$ are exponentials that either simultaneously converge to, or diverge from, the fixed point coordinates x_0 and y_0. The eigenvalues of **A** are real and have the same sign.

Saddle: The components $x(t)$ and $y(t)$ are exponentials; whereas one of the components converges to the fixed point, the other diverges from it. The eigenvalues of **A** are real and have opposite signs.

Spiral: The components $x(t)$ and $y(t)$ are exponentially modulated sinusoidal functions and the resulting streamline forms a spiral curve. The eigenvalues of **A** form a complex conjugate pair.

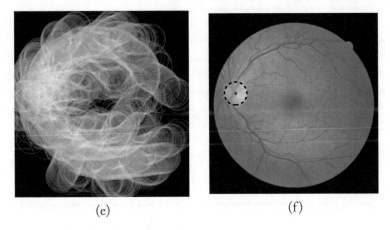

(e)   (f)

Figure 3.11: *Continued.* (a) DRIVE image 01. Image size: 584×565 pixels. (b) Binary edge image using the Sobel operators; the threshold was set to be 0.02. (c) Hough space with $c = 31$ pixels; maximum value = 58. (d) Hough space with $c = 37$ pixels; maximum value = 77. (e) Hough space with $c = 47$ pixels; maximum value = 67. (f) The detected circle with the radius of $c = 37$ pixels centered at (89, 260), superimposed on the original color image. Reproduced with permission from X. Zhu, R.M. Rangayyan, and A.L. Ells "Detection of the optic disc in fundus images of the retina using the Hough transform for circles", *Journal of Digital Imaging*, 23(3): 332-341, June 2010. © Springer.

Table 3.1 lists the three phase portraits and the corresponding orientation fields generated by a system of linear first-order differential equations.

Given an image presenting oriented texture, the orientation field $\theta(x, y)$ of the image is defined as the angle of the texture at each pixel location (x, y). The orientation field of an image can be qualitatively described by the type of phase portrait that is most similar to the orientation field, along with the center of the phase portrait. Such a description can be achieved by estimating the parameters of the phase portrait that minimize the difference between the orientation field of the corresponding phase portrait and the orientation field of the image $\theta(x_i, y_i)$ obtained by Gabor filtering. Let us define x_i and y_i as the x and y coordinates of the ith pixel, $1 \leq i \leq N$. Let us also define $\theta_i = \theta(x_i, y_i)$, and $\phi_i = \phi(x_i, y_i | \mathbf{A}, \mathbf{b})$. The sum of the squared error is given by

$$R(\mathbf{A}, \mathbf{b}) = \sum_{i=1}^{N} \sin^2 (\theta_i - \phi_i) . \tag{3.15}$$

Minimization of $R(\mathbf{A}, \mathbf{b})$ leads to the optimal phase portrait parameters that describe the orientation field of the image under analysis.

In the present work, the matrices $\mathbf{A}$ and $\mathbf{b}$ were estimated using simulated annealing followed by the nonlinear least squares optimization technique [67]. Constraints were placed so that the

Table 3.1: Phase Portraits for a System of Linear First-order Differential Equations [66]. Solid lines indicate the movement of the $p(t)$ and the $q(t)$ components of the solution; dashed lines indicate the streamlines. Reproduced with permission from F.J. Ayres and R.M. Rangayyan, "Characterization of architectural distortion in mammograms via analysis of oriented texture", *IEEE Engineering in Medicine and Biology Magazine,*, 24(1): 59-67, January 2005. © IEEE.

Phase portrait type	Eigenvalues	Streamlines	Appearance of the orientation field
Node	Real eigenvalues of same sign	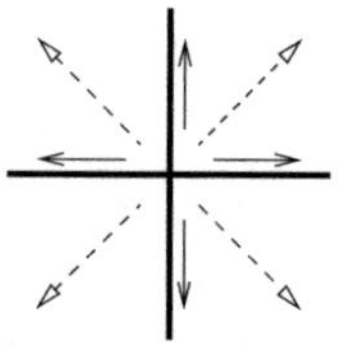	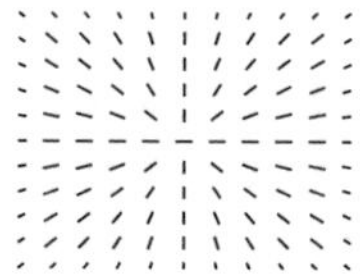
Saddle	Real eigenvalues of opposite sign	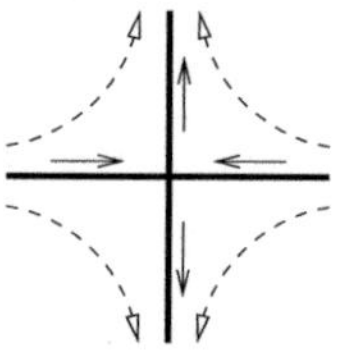	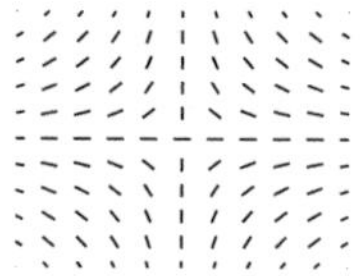
Spiral	Complex eigenvalues	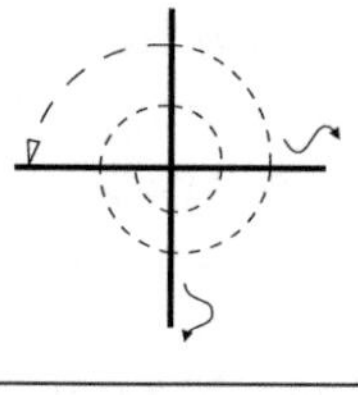	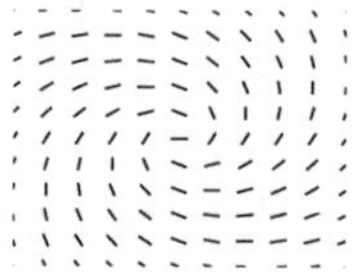

matrix in the phase portrait model is symmetric and has a condition number less than 3.0 [67]. The constrained method yields only two phase portrait maps: node and saddle.

Several patterns of phase portraits may exist in a given image, resulting in the presence of multiple focal or fixed points. The approximation of real orientation fields using Equation 3.13 is valid only locally. A general strategy to extend the method for the local analysis of orientation fields to the analysis of large orientation fields is to analyze the large orientation field at multiple locations, and to accumulate the obtained information in a form that permits the identification of the various relevant structures present in the overall orientation field.

Rao and Jain [66] proposed the following method for the analysis of large orientation fields:

- Create three images of the same resolution as that of the image. These three images will be referred to as the phase portrait maps of the node, saddle, and spiral type. Initialize the phase portrait maps to zero. In the present work, only two maps are used (node and saddle).

- Move a sliding analysis window throughout the orientation field. For every position of the analysis window, find the optimal parameters $\mathbf{A_{opt}}$ and $\mathbf{b_{opt}}$ that best describe the orientation field, and determine the type of phase portrait. Find the fixed-point location associated with the orientation field under the analysis window. Select the phase portrait map corresponding to the phase portrait type determined above and increment the value present at the pixel nearest to the fixed-point location. This procedure is referred to as vote casting.

After all votes are cast, the phase portrait maps may be analyzed to detect the presence of patterns in the given image or orientation field. If a portion of the orientation field is comprised of orientations radiating from a central point, it is expected that the node map will contain a large number of votes close to the geometrical focus of the observed pattern.

The node map of Figure 3.5 (a) is shown in part (d) of the same figure. The locations where the tree branches off have high intensity values in the node map, indicating the presence of a node pattern.

Figure 3.6 (d) and (e) show the node map and the saddle map for the image shown in part (a) of the same figure. The intensity values in each map correspond to the number of votes accumulated at each location, enhanced by using the log operation, and mapped to the range [0, 255] for display purposes. The orientation field obtained using the Gabor filters was downsampled by a factor of 4, and an analysis window of size 40×40 pixels (approximately, twice the average size of the ONH) was slid pixel by pixel through the orientation field. The node map was filtered with a Gaussian filter of standard deviation 6 pixels. Two prominent peaks can be observed in the node map corresponding to intersections of the blood vessels; the largest peak in the node map corresponds to the center of the ONH because the orientation field around the ONH area is similar to a node pattern. The saddle map is not of interest in the present work. Part (f) of the same figure shows the position of the first peak in the node map, located at (76, 252), superimposed on the original image.

3.5   REMARKS

Several approaches to detect edges and geometrical patterns in digital images were reviewed in this chapter. Mean and median filters as well as morphological filters for removing noise and artifacts in images were introduced. Methods for edge detection, including the Sobel and Canny methods, as well as their advantages and disadvantages, were discussed. Matched filters for detecting structures of known characteristics were described. The Gabor filter, a particular type of a matched filter, was discussed in detail for the detection of piecewise-linear structures existing in images of the retina. Regarding the detection of geometrical patterns in an image, the Hough transform for the detection of lines and circles, as well as phase portraits, were studied. The details regarding the application of the above-mentioned methods for processing images of the retina are presented in the following chapters.

CHAPTER 4

Datasets and Experimental Setup

In this chapter, details about the two publicly available datasets of images of the retina, the DRIVE [8, 9] and the STARE [15, 16, 17] datasets, are given. In addition, details of the experimental setup, annotation of the images, and measures of detection performance are provided.

4.1 THE DRIVE DATASET

The images in the DRIVE dataset were acquired using a Canon CR non-mydriatic camera with three charge-coupled device (CCD) detectors and an FOV of 45 degrees. The dataset consists of a total of 40 color fundus photographs. Each image is of size 584 × 565 pixels, represented using 24 bits per pixel in the standard RGB format. Considering the size and FOV of the images, they are low-resolution fundus images of the retina, having an approximate spatial resolution 20 μm per pixel [28].

The DRIVE photographs were obtained from a diabetic retinopathy screening program in the Netherlands. The screening population consisted of 453 subjects between 31 and 86 years of age. Each image was compressed using the JPEG (Joint Photographic Experts Group) procedure, which is common practice in screening programs. Out of the 40 images provided in the DRIVE dataset, 33 are normal and seven contain signs of diabetic retinopathy, namely exudates, hemorrhages, and pigment epithelium changes [8, 9]. The results of manual segmentation of the blood vessels are provided for all 40 images in the dataset; however, no information on the ONH is available.

4.2 THE STARE DATASET

The STARE images were captured using a TopCon TRV-50 fundus camera with an FOV of 35 degrees. Each image is of size 700 × 605 pixels, with 24 bits per pixel; the images have been clipped at the top and bottom of the FOV. No information is available on the spatial resolution of the images. According to Dr. V. Hoover, one of the developers of the STARE dataset, fundus images of the retina do not have the same spatial resolution across the FOV because the interior of the eye being imaged is semi-spherical, like a bowl.

The STARE project was conceived and initiated in 1975 by Michael Goldbaum, M.D., at the University of California, San Diego. Images and clinical data were provided by the Shiley Eye Center at the University of California, San Diego, and by the Veterans Administration Medical

Center in San Diego. A subset of the STARE dataset, which contains 81 images [16], was used in the present work. Out of the 81 images, 30 have normal architecture and 51 have various types of pathology, containing diverse symptoms such as tortuous vessels, choroidal neovascularization, and hemorrhages [16, 17]. This dataset is considered to be more difficult to analyze than the DRIVE dataset because of the inclusion of a larger number of images with pathology of several types. Because the images were scanned from film rather than acquired directly using a digital fundus camera, the quality of the images is poorer than that of the images in the DRIVE dataset. The results of manual segmentation of the blood vessels for a subset of the 81 STARE images are available; however, no information on the ONH is available.

4.3 SCALE FACTOR FOR CONVERTING BETWEEN THE TWO DATASETS

The FOVs of the images in the DRIVE and STARE datasets are different, and hence we cannot use the same spatial resolution. A scale factor provided by ter Haar [27] is used in the present work to convert the spatial resolution between the two datasets. The scale factor is calculated as

$$scale = \frac{\tan(45/2)}{\tan(35/2)} \approx 0.75. \tag{4.1}$$

This equation is based on the geometry of the FOV, as shown in Figure 4.1, and provides an approximate resolution of 15 μm per pixel for the STARE images.

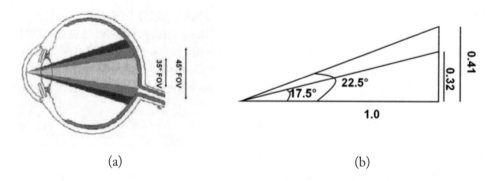

(a)   (b)

Figure 4.1: (a) The area of the retina captured in a photograph in relation to different FOVs. (b) The computation of the scale factor according to the geometry of the FOV. Reproduced with permission from Frank ter Haar, "Automatic localization of the optic disc in digital color images of the human retina", Master's thesis, Utrecht University, Utrecht, the Netherlands, 2005.

4.4   ANNOTATION OF IMAGES OF THE RETINA

In the present work, the performance of each described method is evaluated by comparing the detected center and boundary of the ONH with the same as marked independently by an ophthalmologist and retina specialist (A.L.E.). The center and the contour of the ONH were drawn on each image, by magnifying the original image by 300% using the software ImageJ [69]. When drawing the contour of the ONH, attention was paid so as to avoid the rim of the sclera (scleral cresent or peripapillary atrophy) and the rim of the optic cup, which in some images may be difficult to differentiate from the ONH. When labeling the center of the ONH, care was taken not to mark the center of the optic cup or the focal point of convergence of the central retinal vein and artery .

Figure 4.2 shows the difference between the optic cup and the ONH.

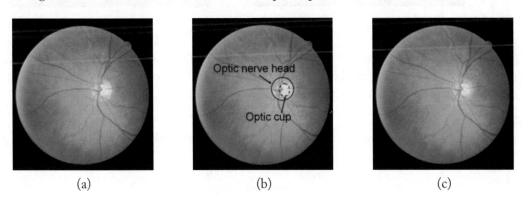

(a)   (b)   (c)

Figure 4.2: (a) Image 04 of the DRIVE dataset. (b) The circle in solid line approximates the boundary of the ONH and the dashed ellipse approximates the boundary of the optic cup. (c) Contour of the ONH marked by an ophthalmologist (A.L.E.).

Several images in the STARE dataset do not contain the full region of the ONH. In such cases, the contour of the ONH was drawn only for the portion lying within the effective region of the image. The mask obtained in the preprocessing step (see Section 3.1) was used to get the part of the contour on the edge of the effective region. Figure 4.3 shows an example with incomplete ONH. In one of the STARE images, the center of the ONH is located outside the FOV. In this case, the center was marked at the estimated location outside the effective region.

4.5   EVALUATION OF THE RESULTS OF DETECTION

4.5.1   MEASURES OF DISTANCE AND OVERLAP

The results of detection of the ONH consist of the center point and the radius of a circle (in the case of the method based on the Hough transform). In order to evaluate the accuracy of the results, for each image, the Euclidean distance between the detected center and the corresponding center marked by the ophthalmologist was computed in pixels and converted to mm. In addition, the overlap

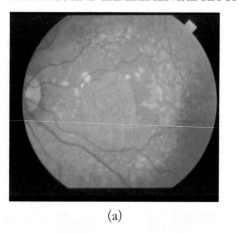
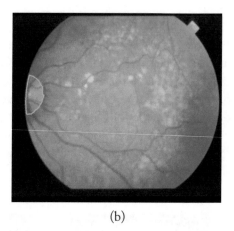

(a)   (b)

Figure 4.3: (a) Image im0003 of the STARE dataset. (b) The manually marked center and contour of the ONH. The mask obtained in the preprocessing step (see Section 3.1) was used to obtain the part of the contour on the edge of the effective region.

between the circular approximation and the contour of the ONH drawn by the ophthalmologist was computed as

$$overlap = \frac{A \cap B}{A \cup B}, \tag{4.2}$$

where A is the region marked by the ophthalmologist and B is the region detected by the described method; see Figure 4.4. The value of overlap is limited to the range $[0, 1]$.

For the images in the STARE dataset which do not contain the complete ONH, the regions used to compute the overlap were limited to their parts lying within the corresponding effective regions of the fundus images, determined by applying the mask generated for each image as described in Section 3.1.

4.5.2   FREE-RESPONSE RECEIVER OPERATING CHARACTERISTICS

Free-response receiver operating characteristics (FROC) provide details of the variation of the sensitivity of detection with respect to the mean number of false-positive responses per image [70, 71]. FROC analysis is applicable when there is no specific number of true negatives, which is the case in the present study.

For the images in the DRIVE dataset, a result in the present work is considered to be successful if the detected ONH center is positioned within the average radius of 0.8 mm (40 pixels for the DRIVE images) of the manually identified center; otherwise, it is labeled as a false positive. Using the scale factor provided by ter Haar [27] (see Section 4.3), for the images in the STARE dataset, the corresponding distance between the detected ONH and the manually identified center for successful

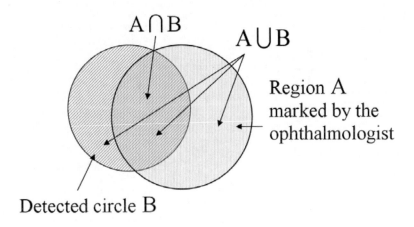

Figure 4.4: Illustration of the measure of overlap between two regions A and B.

detection is 53 pixels. The sensitivity of detection is calculated as the ratio of the number of images with successful detection of the ONH to the number of images analyzed.

The acceptable distance value of 60 pixels, which has been used by other researchers with the STARE dataset [16, 27, 31], when converted using the above-mentioned scale factor, yields a limit of 46 pixels for the images in the DRIVE dataset. In order to facilitate comparison with other published works with the same datasets but different criteria for successful detection, FROC curves were derived using all of the limits mentioned above.

4.6   REMARKS

The DRIVE and STARE datasets used for evaluation of the methods were described in this chapter. An ophthalmologist (A.L.E.) marked the center and contour of the ONH in the images; the two datasets do not provide such information. The methods used for evaluation of the results obtained, including measures of distance and overlap as well as FROC, were described. The results obtained with the datasets and evaluation methods are presented in Chapter 5 and Chapter 6.

CHAPTER 5

Detection of the Optic Nerve Head Using the Hough Transform

The procedure designed and developed in the present work to detect the ONH using the Hough transform is summarized in the flowchart in Figure 5.1. Because the ONH usually appears as a circular region, an algorithm for the detection of circles may be expected to locate the ONH in a retinal fundus image [72]. In the present work, the Hough transform for the detection of circles is used. Before applying the Hough transform, methods for detecting edges are applied to obtain an edge map, which is used as the input to the Hough transform. The best-fitting circle for the ONH is chosen by using a method of intensity-based selection. The sections of the book that provide descriptions of the various steps are labeled in Figure 5.1. The step of selection of the best-fitting circular approximation of the ONH using the reference intensity is described in the present chapter.

5.1 DERIVATION OF THE EDGE MAP

In the present work, the edge map is obtained by applying the Sobel or Canny method to the preprocessed image followed by thresholding, as described in Section 3.2. If the threshold is set too low, there will be a large number of nonzero pixels in the edge map, leading to more computation with the Hough transform. If the threshold is too high, there could be very few nonzero pixels to define an approximate circle in the ONH area, which could cause the detection method to fail. An optimal threshold may need to be determined for each dataset. For the images in the DRIVE dataset, the edge images obtained using the Sobel operators were binarized using a fixed threshold of 0.02; the threshold for the results of the Canny method was fixed at 0.17. For the images in the STARE dataset, the edge images obtained using the Sobel operators were binarized using a threshold chosen automatically by the MATLAB [58] version of the Sobel method; the threshold for the results of the Canny method was fixed at 0.17. The thresholds were chosen by experimentation with a number of images from the DRIVE and STARE datasets, involving comparison of the edge maps obtained at various thresholds with the blood vessels visible in the original images.

The edge maps obtained for the image 01 from the DRIVE dataset are shown in Figure 5.2. The edge maps for the image im0001 from the STARE dataset are shown in Figure 5.3. We can observe that, in the case of the DRIVE image in Figure 5.2, edges can be found around the

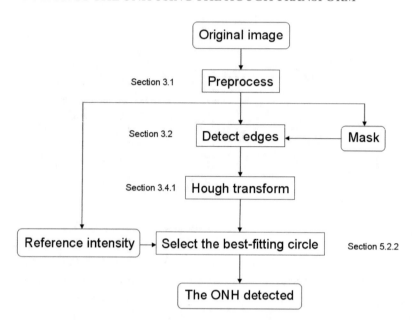

Figure 5.1: Flowchart of the procedure developed to detect the ONH using the Hough transform. The sections of the book that provide descriptions of the various steps are labeled. The step of selection the best-fitting circular approximation of the ONH using the reference intensity is described in the present chapter.

boundary of the ONH with both the Canny and Sobel methods. The Canny method gives connected edges whereas the Sobel operators give disconnected edges. The edge map resulting from the Canny method is much cleaner than that from the Sobel operators, without a large number of isolated points related to the small vessels present in the image. However, the Canny method is computationally more complex than the Sobel operators because of its optimization procedures. In the case of the STARE image in Figure 5.3, we can hardly find edges around the boundary of the ONH because the ONH is obscured by the effects of pathology.

5.2   ANALYSIS OF THE HOUGH SPACE

5.2.1   PROCEDURE FOR THE DETECTION OF THE ONH

After obtaining the edge map, the Hough transform was used to detect the circles existing in the image (see Section 3.4.1). The Hough accumulator is a 3D array, each cell of which is incremented for each nonzero pixel of the edge map that meets the stated condition. For example, the value for

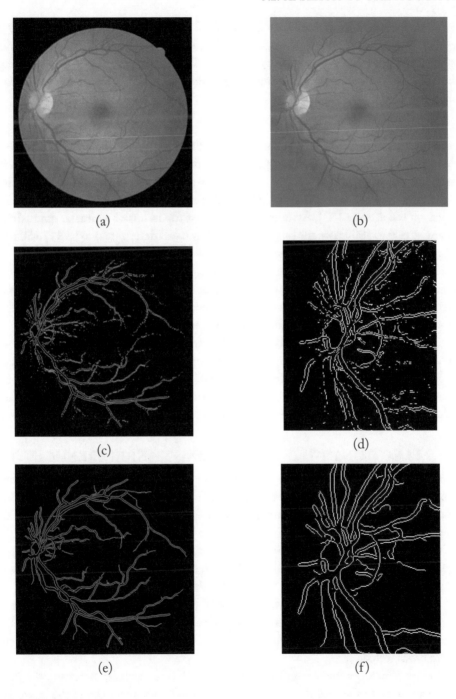

Figure 5.2: Caption on the next page.

Figure 5.2: (a) Image 01 of the DRIVE dataset. (b) Preprocessed luminance image. (c) Binary edge image obtained using the output of the MATLAB version of the Sobel operators thresholded at 0.02 of the normalized intensity. (d) Part (c) magnified in the ONH area. (e) Binary edge image using the output of the MATLAB version of the Canny method thresholded at 0.17 of the normalized intensity. (f) Part (e) magnified in the ONH area.

the cell (a, b, c) in the Hough accumulator is equal to the number of edge map pixels of a potential circle in the image with the center at (a, b) and radius c. In the case of the images in the DRIVE dataset, the size of each image is 584×565 pixels (see Section 4.1). The spatial resolution of the images in the DRIVE dataset is about 20 μm per pixel. The physical diameter of the ONH is about 1.5 mm, on the average [32]. Assuming the range of the radius of a circular approximation to the ONH to be 600 to 1000 μm, the range for the radius c was determined to be 31 to 50 pixels. Hence, the size of the Hough accumulator was set to be $584 \times 565 \times 20$.

For the STARE images, the size of the images is 700×605 pixels (see Section 4.2). The approximate spatial resolution is 15 μm per pixel, obtained by converting that of the DRIVE images with the scale factor derived by ter Haar [27] (see Section 4.3). Hence, the range for c was determined to be 46 to 65 pixels (limited to 20 pixels) for images from the STARE dataset.

The potential circles indicated by the Hough accumulator were ranked in terms of the corresponding value of the accumulator cell, and the top 30 were selected for further analysis. The top three circles in the Hough space for the test image 02 in the DRIVE dataset are shown in Figure 5.4. We can observe that the first circle in the Hough space fits the actual ONH well, whereas the second and third circles match arches of the blood vessels.

5.2.2   SELECTION OF THE CIRCLE USING THE REFERENCE INTENSITY

The first circle indicated by the Hough accumulator may not always correspond to the best circular approximation of the ONH. To address this limitation, a criterion was included to select the most appropriate circle by using the intensity information, shown as the last step in the flowchart in Figure 5.1.

After preprocessing (see Section 3.1), a 5×5 median filter was applied to the luminance image, to remove outliers (noisy pixels) in the image. Then, the maximum intensity in each image was calculated to serve as a reference intensity for the selection of circles.

Because we expect the ONH to be one of the bright areas in the fundus image, a threshold equal to 0.9 times the reference intensity was used to check the maximum intensity within a circular area with half of the radius of each potential circle. If the test failed, the circle was rejected, and the next circle was tested.

Figure 5.5 shows two examples from the DRIVE dataset. In each case, the blue dash-dot circle corresponds to the global maximum in the Hough parameter space; the black dashed circle corresponds to the highest local maximum in the Hough space that also meets the condition based

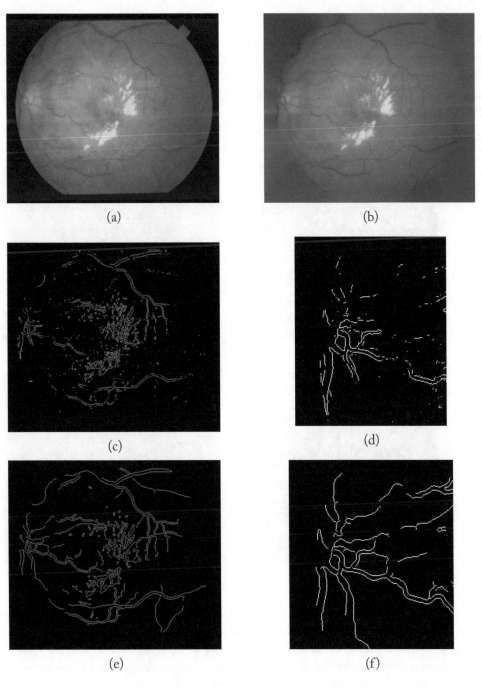

Figure 5.3: Caption on the next page.

Figure 5.3: (a) Image im0001 of the STARE dataset. (b) Preprocessed luminance image. (c) Binary edge image obtained using the output of the MATLAB version of the Sobel operators with an automatically chosen threshold. (d) Part (c) magnified in the ONH area. (e) Binary edge image obtained using the output of the MATLAB version of the Canny method thresholded at 0.17 of the normalized intensity. (f) Part (e) magnified in the ONH area.

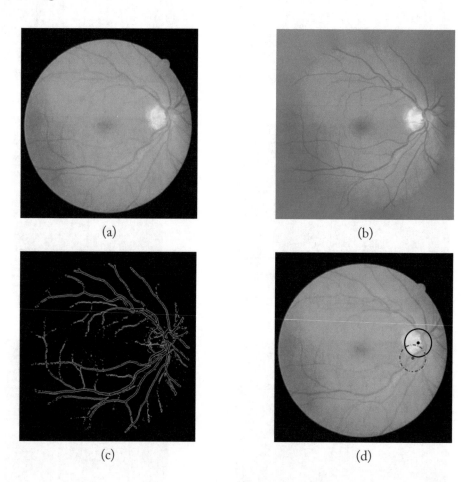

Figure 5.4: (a) Image 02 of the DRIVE dataset. (b) Preprocessed luminance image. (c) Binary edge image obtained using the MATLAB version of the Sobel operators thresholded at 0.02 of the normalized intensity. (d) The top three circles in the Hough space superimposed on the original color image. The black contour in solid line corresponds to the first circle. The cyan dashed circle corresponds to the second circle. The blue dash-dot circle corresponds to the third circle.

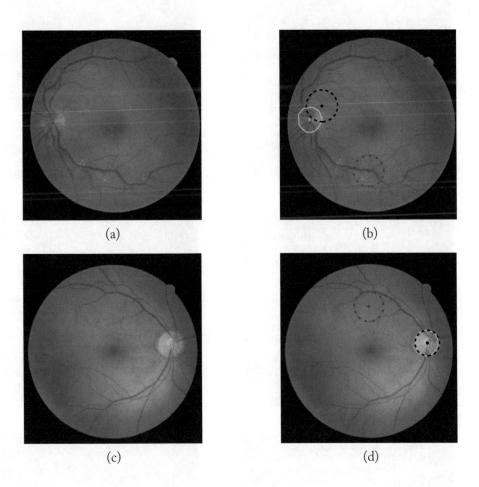

(a)   (b)

(c)   (d)

Figure 5.5: (a) Test image 03 of the DRIVE dataset. (b) The result of detection of the ONH for image 03. The blue dash-dot circle corresponds to the global maximum in the Hough parameter space. The black dashed circle corresponds to the highest local maximum in the Hough space that also meets the condition based on 90% of the reference intensity. The cyan contour in solid line is the contour of the ONH marked by the ophthalmologist. Distance = 1.15 mm, overlap = 0.12 after using the intensity condition. (c) Test image 10 of the DRIVE dataset. (d) Result for image 10 with distance = 0.04 mm, overlap = 0.90 after using the intensity condition. Reproduced with permission from X. Zhu, R.M. Rangayyan, and A.L. Ells "Detection of the optic disc in fundus images of the retina using the Hough transform for circles", *Journal of Digital Imaging,* 23(3): 332-341, June 2010. © Springer.

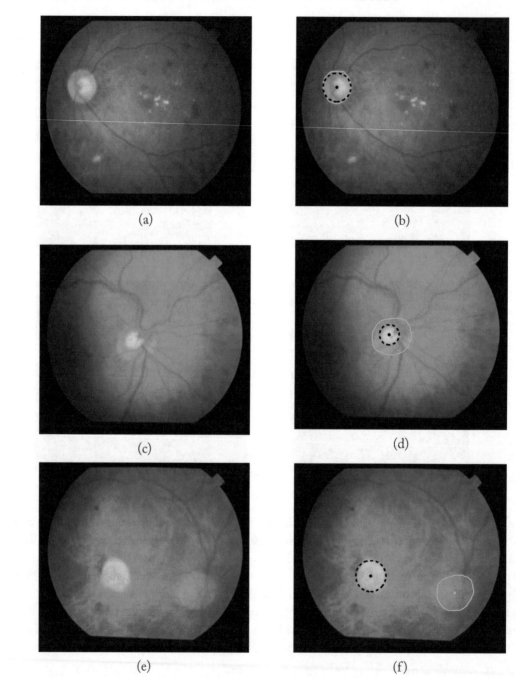

(a)

(b)

(c)

(d)

(e)

(f)

Figure 5.6: Caption on the next page.

Figure 5.6: (a) Image im0009 of the STARE dataset. (b) The result of detection of the ONH for image im0009. The black dashed circle corresponds to the highest local maximum in the Hough space that also meets the condition based on 90% of the reference intensity. The cyan contour in solid line is the contour of the ONH marked by the ophthalmologist. Distance = 3 pixels, overlap = 0.87 after using the intensity condition. (c) Image im0030. (d) Result for image im0030 with distance = 9 pixels, overlap = 0.26 after using the intensity condition. (e) Image im0036. (f) Result for image im0036 with distance = 281 pixels, overlap = 0 after using the intensity condition.

on 90% of the reference intensity; the cyan contour in solid line is the contour of the ONH marked by the ophthalmologist. In the case of image 03, the intensity-based selection step rejected the first circle indicated by the Hough transform, which is not related to the ONH; the finally selected circle contains a part of the bright area in the image corresponding to the ONH and could not be rejected, even though it does not fit the ONH. In the case of image 10, the selection method was successful; the result increased the overlap and decreased the distance between the marked and detected contours and centers of the ONH, with distance = 1.15 mm, overlap = 0.12 after using the intensity condition.

Figure 5.6 shows three examples from the STARE dataset. We can observe in part (e) of the figure that it is not always the case that the ONH is the brightest area in fundus images, especially in images of pathological cases in the STARE dataset.

5.3   RESULTS OF DETECTION OF THE OPTIC NERVE HEAD

The results of the evaluation of detection of the ONH for the DRIVE dataset are shown in Table 5.1. The mean distance of the detected center of the ONH with the intensity-based condition is 0.36 mm (18 pixels); the average overlap is 0.73.

For FROC analysis, the top 10 potential circles in the Hough space were selected in order to test for the detection of the ONH. The FROC curve for the DRIVE dataset is shown in Figure 5.7, which indicates a sensitivity of 92.5% at 8.9 false positives per image; note that the intensity-based condition for the selection of circles is not applicable in FROC analysis.

Table 5.2 gives the evaluation of the results for the STARE dataset, indicating an average overlap of only 0.32. The FROC curve for the STARE dataset is shown in Figure 5.8; a sensitivity of 70.4% was obtained at 9 false positives per image.

5.4   DISCUSSION

In Table 5.3, the success rates of locating the ONH reported by several methods published in the literature and reviewed in Chapter 2 are listed. However, there is no established standard or common criterion for successful detection. Some of the published reports are not clear about how a successful

Table 5.1: Statistics of the distance and overlap between the manually marked and the detected ONH for the 40 images in the DRIVE dataset. Min = minimum, max = maximum, std = standard deviation. Reproduced with permission from X. Zhu, R.M. Rangayyan, and A.L. Ells "Detection of the optic disc in fundus images of the retina using the Hough transform for circles", *Journal of Digital Imaging*, 23(3): 332-341, June 2010. © Springer.

Method	Distance mm (pixels)				Overlap			
	mean	min	max	std	mean	min	max	std
First peak in Hough space	1.05 (52.5)	0.03 (1.5)	7.53 (376.5)	1.87 (93.5)	0.58	0	0.95	0.36
Peak selected using intensity condition	0.36 (18)	0.02 (1.0)	6.17 (308.5)	1.00 (50)	0.73	0	0.95	0.25

Table 5.2: Statistics of the distance and overlap between the manually marked and the detected ONH for the 81 images from the STARE dataset. Min = minimum, max = maximum, std = standard deviation.

Method	Distance (pixels)				Overlap			
	mean	min	max	std	mean	min	max	std
First peak in Hough space	150.5	1.0	469	140.5	0.21	0	0.91	0.27
Peak selected using intensity condition	132.5	1.0	527	159	0.32	0	0.91	0.30

detection was defined. Some researchers used their own datasets instead of the publicly available datasets, which makes comparative analysis difficult.

The method described in the present chapter was tested on two publicly available datasets. In Table 5.4, the various methods used to evaluate the detection of the ONH with the DRIVE dataset are listed. It is not clear whether the manual marking of the ONH was done by an ophthalmologist in the other works listed. The present method has been evaluated with the center and contour of the ONH marked independently by an ophthalmologist (A.L. E.). Although the success rate of the method described in the present chapter with the DRIVE dataset is not the highest among the works listed in the table, the method has the advantage that it does not require preliminary detection of blood vessels, and hence has lower complexity than some of the other methods reported. The difference of one pixel between the average distance measure of the results of Youssif et al. [31] and the method described in the present chapter is negligible. Similar methods reported by Barrett

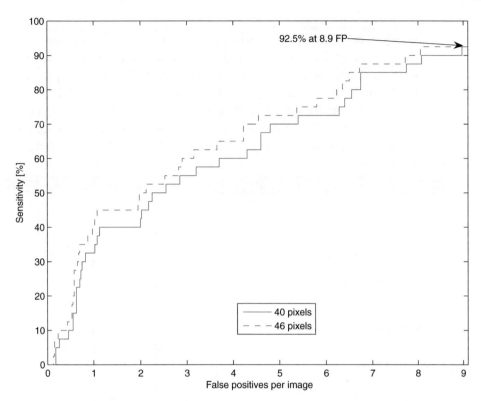

Figure 5.7: FROC curves for the DRIVE dataset (40 images). The dashed line corresponds to the FROC curve with a threshold for successful detection of 46 pixels from the manually marked center to the detected center. The solid line is the FROC obtained when the threshold is set to be 40 pixels. FP = false positive. Reproduced with permission from X. Zhu, R.M. Rangayyan, and A.L. Ells "Detection of the optic disc in fundus images of the retina using the Hough transform for circles", *Journal of Digital Imaging*, 23(3): 332-341, June 2010. © Springer.

et al. [26], ter Haar [27], and Chrástek et al. [28, 29] were not tested with the publicly available DRIVE dataset to facilitate comparative analysis. The method described in the present chapter can locate the center of the ONH and provide a circular approximation to its boundary. However, the method may fail when the ONH is dim or blurred because it is based on the expected property that the ONH is one of the bright areas in the image with an approximately circular boundary.

A comparative analysis of the Sobel and the Canny methods for edge detection was conducted. It was found that the Sobel operators are more suitable for the present application. The Canny method gives complicated and connected edges, whereas the Sobel operators give simple and disconnected edges. The Hough transform can detect objects of the specified geometry with disconnected nonzero edge pixels; the ONH can be detected efficiently from the relatively sparse edge map obtained using

Table 5.3: Comparison of the efficiency of locating the ONH in images of the retina obtained by different methods. For the images of the DRIVE dataset, a result is considered to be successful if the detected ONH center is positioned within 46 pixels of the manually identified center. For the STARE dataset, the threshold is 60 pixels. Reproduced with permission from X. Zhu, R.M. Rangayyan, and A.L. Ells "Detection of the optic disc in fundus images of the retina using the Hough transform for circles", *Journal of Digital Imaging*, 23(3): 332–341, June 2010. © Springer.

Method of detection	DRIVE	STARE	Other dataset
Lalonde et al. [32]	-	-	93%
Sinthanayothin et al. [23]	-	-	99.1%
Osareh et al. [30]	-	-	90.3%
Hoover and Goldbaum [16]	-	89%	-
Foracchia et al. [33]	-	97.5%	-
ter Haar [27]	-	93.8%	-
Park et al. [25]	90.3%	-	-
Ying et al. [34]	97.5%	-	-
Kim et al. [35]	-	-	91%
Fleming et al. [13]	-	-	98.4%
Park et al. [36]	-	91.1%	-
Youssif et al. [31]	100%	98.8%	-
Method using the Hough transform as described in the present chapter [72]	90%	44.4%	-

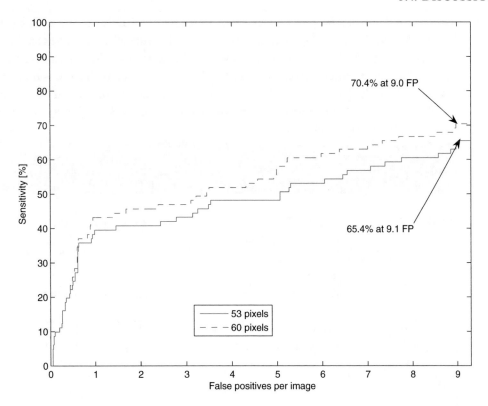

Figure 5.8: FROC curves for the STARE dataset (81 images). The dashed line corresponds to the FROC curve with a threshold for successful detection of 60 pixels from the manually marked center to the detected center. The solid line is the FROC obtained when the threshold is set to be 53 pixels. FP = false positive.

the Sobel operators. On the other hand, vessel arches present in the edge map obtained using the Canny method caused false detection. Furthermore, the complicated edge maps with more nonzero pixels obtained using the Canny method led to more computational requirement in the Hough space than the case with the Sobel operators. In a preliminary study [73], the rate of detection of the ONH obtained using the edge maps provided by the Sobel gradients was 92.50% with the 40 images from the DRIVE dataset. The corresponding rate using the edge maps provided by the Canny method was lower, at 80%. Based on this result, subsequent work, as described in the present chapter, was limited to the use of the Sobel gradient.

The analysis of the results of detection of the ONH was performed in two different ways. One approach involved the assessment of the overlap and distance between the manually marked and detected ONHs. The second approach was based on FROC analysis, which has not been reported in any of the published works on the detection of the ONH.

Table 5.4: Comparison of different methods tested with the DRIVE dataset, in terms of the distance between the manually marked and the detected center. The definition of success is specified as the maximum allowed distance between the manually marked and the detected center for a successful detection. Reproduced with permission from X. Zhu, R.M. Rangayyan, and A.L. Ells "Detection of the optic disc in fundus images of the retina using the Hough transform for circles", *Journal of Digital Imaging*, 23(3): 332-341, June 2010. © Springer.

Authors	Method	Manual marking	Average distance (pixels)	Definition of success
Park et al. [25]	Brightness and Hough transform	yes	not provided	not specified
Youssif et al. [31]	Direction matched filter	yes	17	60 pixels
Ying et al. [34]	Brightness and local fractal analysis	no	not provided	not specified
The method as described in the present chapter [72]	Brightness and Hough transform	yes	18	46 pixels

From Table 5.1 and Table 5.2, we can find that, with the inclusion of selection based on the reference intensity, the average distance was reduced and the average overlap was increased, leading to better performance of the described method with both the DRIVE and STARE datasets. However, in the case of the STARE dataset, the performance was not improved as much as with the DRIVE dataset after intensity-based selection. This is because the ONH is not always one of the bright areas in the images in the STARE dataset.

The appearance of the ONH in the images in the STARE dataset varies significantly due to various types of retinal pathology. Figure 5.6 (b) gives an example from the STARE dataset where the method described in the present chapter successfully detected the ONH with distance = 3 pixels and overlap = 0.87. Figure 5.6 (d) is an example where the center of the ONH was successfully located, whereas the contour detected was that of the optic cup, which is a smaller and brighter circular region within the ONH. Several images in the STARE dataset are of poor quality: they are out of focus, over exposed, or under exposed. As a result, the method did not yield good results. Misleading features that affected the performance of the method with the STARE dataset were grouped by the ophthalmologist (A.L.E.) into vessel curvature (20 out of 81 images), retinal fibrosis, scleral crescent, and white lesions (including retinal edema, exudates, and drusen; 16 out of 81 images). In addition, the poor quality of three images led to failure.

5.5   REMARKS

The procedure using the Hough transform to locate the ONH was described in detail in the present chapter. Based on the properties of the ONH, the procedure includes edge detection using the Sobel method, and detection of circles using the Hough transform. The Hough transform assists in the detection of the center and radius of a circle that approximates the margin of the ONH. With an intensity-based criterion for the selection of the circles and a limit of 46 pixels (0.92 mm) on the distance between the center of the detected circle and the manually identified center of the ONH, a successful detection rate of 90% was obtained with 40 images from the DRIVE dataset. Analysis of the results was performed in two different ways. One approach involved the assessment of the overlap and distance between the manually identified and automatically detected ONHs, the former independently marked by an ophthalmologist. A mean distance of 0.36 mm (18 pixels) was obtained with the 40 images from the DRIVE dataset. The second approach was based on FROC analysis, which has not been reported in any of the published works on the detection of the ONH. The Hough-transform-based method is compared with another method for the detection of the ONH in the following chapter.

CHAPTER 6

Detection of the Optic Nerve Head Using Phase Portraits

The procedure to detect the center of the ONH using phase portraits is summarized in the flowchart in Figure 6.1. Because the center of the ONH is usually close to the focal point of the central retinal vein and artery, a method is described to detect the center of the ONH via the detection of the blood vessels of the retina using Gabor filters [12, 63] and detection of convergence of the vessels using phase portrait modeling [61, 66, 67, 74]. The best-fitting point for the center of the ONH is chosen by an intensity-based condition. The sections of the book that provide descriptions of the various steps are labeled in Figure 6.1.

6.1   DERIVATION OF THE ORIENTATION FIELD

Gabor filters [55, 62] (see Section 3.3.2) were used to detect blood vessels in the images of the retina. The outputs of Gabor filters are the magnitude response image and the orientation field. Figure 6.2 (a) shows training image 34 from the DRIVE dataset; part (b) of the same figure shows the output of the preprocessing step (see Section 3.1). The image in part (c) of the same figure shows the magnitude response of Gabor filters, using a bank of 180 Gabor filters with $\tau = 8$ pixels and $l = 2.9$ (see Section 3.3.2). The magnitude response image was composed by selecting the maximum response over all of the Gabor filters for each pixel. We can observe that the Gabor filters give a high response in magnitude for most of the vessels. Part (d) of Figure 6.2 shows the orientation field, which is obtained as the angle of the Gabor filter with the largest magnitude response for each pixel. (Needles indicating the local orientation have been drawn for every fifth pixel in the row and column directions.) The mapping of the orientation of the blood vessels can be seen clearly in the orientation field.

6.2   ANALYSIS OF THE NODE MAP

The orientation field was analyzed using phase portraits (see Section 3.4.2). In the present work, an analysis window of size 40×40 pixels was slid pixel by pixel through the orientation field. Constraints were placed so that the matrix in the phase portrait model is symmetric and has a condition number less than 3.0 [67], so as to reject degenerate patterns. The constrained method

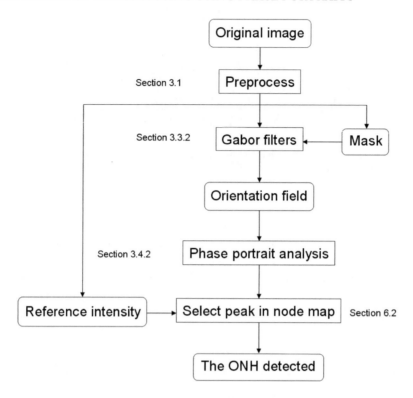

Figure 6.1: Flowchart of the procedure to detect the ONH using phase portraits. The sections of the book that provide descriptions of the various steps are labeled.

yields two phase portrait maps: node and saddle. A vote was cast in the node or saddle map, as indicated by the eigenvalues of the matrix in the phase portrait model, at the location given by the corresponding fixed point. To avoid uncertainty in the fixed point, a condition was placed on the distance between the fixed point and the center of the corresponding analysis window, so as to be not greater than 200 pixels. Because patterns are usually ill-defined at the fixed point, the same distance was also constrained to be more than 5 pixels. The node map was filtered with a Gaussian filter of standard deviation 6 pixels so that multiple peaks do not exist within a circular area with the average radius of the ONH (40 pixels or 0.80 mm for the DRIVE images).

All peaks in the filtered node map were detected and rank-ordered by magnitude. The result was used to label positions related to the sites of convergence of the blood vessels [67]. In Figure 6.3, the positions of the peaks of the node maps are shown for four images from the DRIVE dataset. The number marked is the rank order of the corresponding point in the node map, with no condition imposed regarding intensity in the original image. We can observe that bifurcations of blood vessels lead to high responses in the node map. In all of the four cases, the first peak in the node map indicates

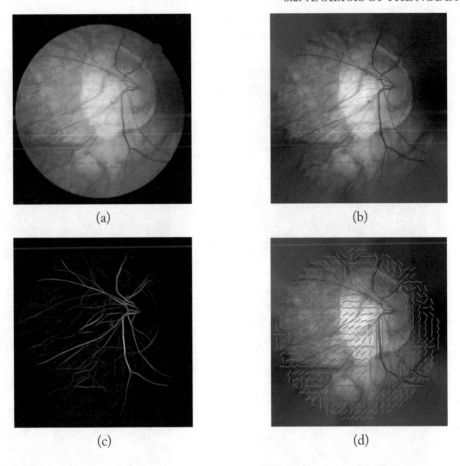

Figure 6.2: (a) DRIVE image 34. (b) Preprocessed luminance image. (c) Magnitude response of the Gabor filters. (d) Filtered and downsampled orientation field. Needles indicating the local orientation have been drawn for every fifth pixel in the row and column directions.

the center of the ONH; however, in general, this is not always the case. To address this problem, an additional step of intensity-based selection of the center of the ONH was used. A circular area was extracted from the Y component with the detected peak location as the center and radius equal to 20 pixels or 0.40 mm, corresponding to about half of the average radius of the ONH (for the DRIVE images). Pixels within the selected area were rank-ordered by their brightness values, and the top 1% were selected. The average of the selected pixels was computed. If the average brightness was greater than 68% of the reference intensity (for the DRIVE images) obtained as described in Section 3.1, the peak location was accepted as the center of the ONH; otherwise, the next peak was checked. For the STARE images, the threshold was set as 0.5 times the reference intensity. The thresholds were chosen by experimentation.

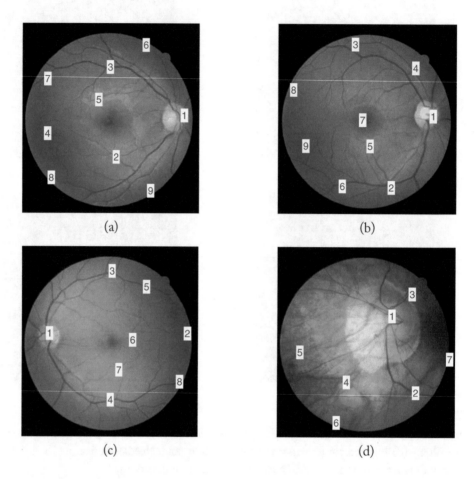

Figure 6.3: (a) Training image 37 from the DRIVE dataset. The number marked is the rank order of the corresponding point in the node map; the positions of all peaks are shown. (b) Training image 36. (c) Training image 35. (d) Training image 34. Reproduced with permission from R.M. Rangayyan, X. Zhu, F.J. Ayres, and A.L. Ells, "Detection of the optic nerve head in fundus images of the retina with Gabor filters and phase portrait analysis", *Journal of Digital Imaging*, 23(4): 438-453, August 2010. © Springer.

In Figure 6.4 (a), STARE image im0021 is shown; the magnitude response image and the orientation field obtained using the Gabor filter are shown in parts (b) and (c) in Figure 6.4. Parts (d) and (e) show the node and saddle map derived from the orientation field using phase portrait analysis. We can observe a strong peak in the node map at the center of the ONH; peaks are seen

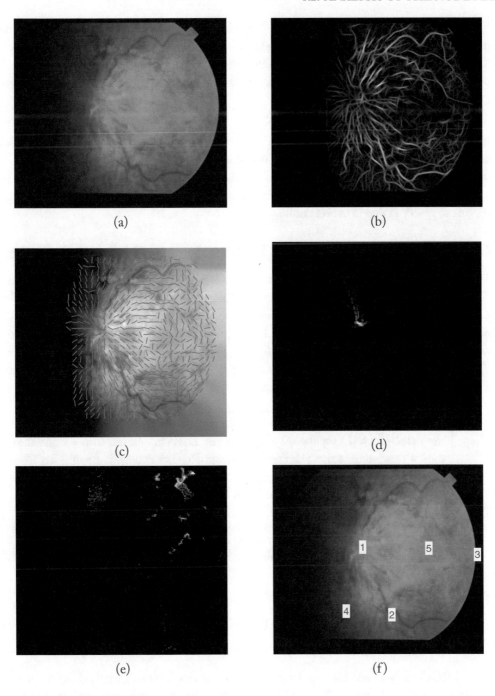

Figure 6.4: Caption on the next page.

Figure 6.4: (a) Image im0021 from the STARE dataset. (b) Magnitude response of the Gabor filters. (c) Filtered and downsampled orientation field. Needles indicating the local orientation have been drawn for every fifth pixel in the row and column directions. (d) Node map. (e) Saddle map. (f) Peaks in the node map which also satisfy the intensity-based condition. Each number marked is the rank order of the corresponding peak in the node map.

in the saddle map around vessel branches. It is evident that the node map can be used to locate the center of the ONH. In part (f) in Figure 6.4, the locations of the top five peaks that also meet the intensity-based condition are shown, superimposed on the original image. The first peak is located at the center of the ONH; the second peak is located at a retinal vessel bifurcation. The peak on the edge of the FOV is due to camera noise.

6.3   RESULTS OF DETECTION OF THE OPTIC NERVE HEAD

The method described in this chapter gives the center of the ONH; therefore, the Euclidean distance between the manually marked and the detected center of the ONH was used to evaluate the results. The statistics of distance between the manually marked and the detected center of the ONH for the DRIVE images are shown in Table 6.1. The mean distance of the detected center of the ONH with the intensity-based condition is 0.46 mm (23.2 pixels). The statistics of the distance for the STARE images are shown in Table 6.2; the statistics are the same with or without the intensity-based condition, with a mean distance of 119 pixels.

Table 6.1: Statistics of the Euclidean distance between the manually marked and the detected ONH for the 40 images in the DRIVE dataset. Min = minimum, max = maximum, std = standard deviation. Reproduced with permission from R.M. Rangayyan, X. Zhu, F.J. Ayres, and A.L. Ells, "Detection of the optic nerve head in fundus images of the retina with Gabor filters and phase portrait analysis", *Journal of Digital Imaging*, 23(4): 438-453, August 2010. © Springer.

Method	Distance mm (pixels)			
	mean	min	max	std
First peak in node map	1.61 (80.7)	0.03 (1.4)	8.78 (439)	2.40 (120)
Peak selected using intensity condition	0.46 (23.2)	0.03 (1.4)	0.91 (45.5)	0.21 (10.4)

FROC analysis was also used for evaluation of the described method. In order to evaluate the detection of the center of the ONH, if more than 10 peaks were present in the node map, the top 10 peaks in the node map were selected; otherwise, all of the peaks in the node map were used.

Table 6.2: Statistics of the Euclidean distance between the manually marked and the detected ONH for the 81 images from the STARE dataset. Min = minimum, max = maximum, std = standard deviation. Reproduced with permission from R.M. Rangayyan, X. Zhu, F.J. Ayres, and A.L. Ells, "Detection of the optic nerve head in fundus images of the retina with Gabor filters and phase portrait analysis", *Journal of Digital Imaging*, 23(4): 438-453, August 2010. © Springer.

Method	Distance (pixels)			
	mean	min	max	std
First peak in node map	119	1.4	544	156
Peak selected using intensity condition	119	1.4	544	156

The FROC curves for both the DRIVE and STARE datasets were prepared with two definitions of successful detection (as described in Section 4.5.2). The FROC curves for the DRIVE images are shown in Figure 6.5, which indicates a sensitivity of 100% at 2.65 false positives per image with the distance limit of 46 pixels. The intensity-based condition for the selection of centers is not applicable in FROC analysis. The FROC curves for the STARE images are shown in Figure 6.6; a sensitivity of 88.9% was obtained at 4.6 false positives per image with the distance limit of 60 pixels.

6.4 DISCUSSION

In Table 6.3, the success rates of locating the ONH reported by several methods published in the literature and reviewed in Chapter 2 are listed. The center of the ONH was successfully detected in all of the 40 images in the DRIVE dataset by the method based on phase portraits. The distance for successful detection used in the present work is 46 pixels, whereas the other methods listed used 60 pixels [16, 31]. The method using phase portraits has performed better than some of the recently published methods with images from the DRIVE dataset.

The appearance of the ONH in the images in the STARE dataset varies significantly due to various types of retinal pathology. Misleading features that affected the performance of the method using phase portraits with the STARE dataset were grouped by the ophthalmologist (A.L.E.) as alternate retinal vessel bifurcation (12 out of 81 images), convergence of small retinal vessels at the macula (5 out of 81 images), and camera noise (2 out of 81 images). Figure 6.7 gives four examples from the DRIVE dataset where the phase portraits algorithm resulted in a peak in the macular region. The numbers marked on the images are the ranks of the corresponding peaks in the node map. (Only the peaks on or near the ONH and macula are shown.) In the case of test image 06 and training image 40, shown in parts (a) and (d) in Figure 6.7, the peak in the macular region ranked second among the peaks in the node map. In the other two cases, shown in parts (b) and (c) in

Table 6.3: Comparison of the efficiency of locating the ONH in images of the retina obtained by different methods. For the images of the DRIVE dataset, a result is considered to be successful if the detected ONH center is positioned within 46 pixels of the manually identified center. For the STARE dataset, the threshold is 60 pixels. Reproduced with permission from R.M. Rangayyan, X. Zhu, F.J. Ayres, and A.L. Ells, "Detection of the optic nerve head in fundus images of the retina with Gabor filters and phase portrait analysis", *Journal of Digital Imaging*, 23(4): 438-453, August 2010. © Springer.

Method of detection	DRIVE	STARE	Other dataset
Lalonde et al. [32]	-	-	93%
Sinthanayothin et al. [23]	-	-	99.1%
Osareh et al. [30]	-	-	90.3%
Hoover and Goldbaum [16]	-	89%	-
Foracchia et al. [33]	-	97.5%	-
ter Haar [27]	-	93.8%	-
Park et al. [25]	90.3%	-	-
Ying et al. [34]	97.5%	-	-
Kim et al. [35]	-	-	91%
Fleming et al. [13]	-	-	98.4%
Park et al. [36]	-	91.1%	-
Youssif et al. [31]	100%	98.8%	-
The present work using Hough transform [72]	90%	44.4%	-
The present work using phase portraits [74]	100%	69.1%	-

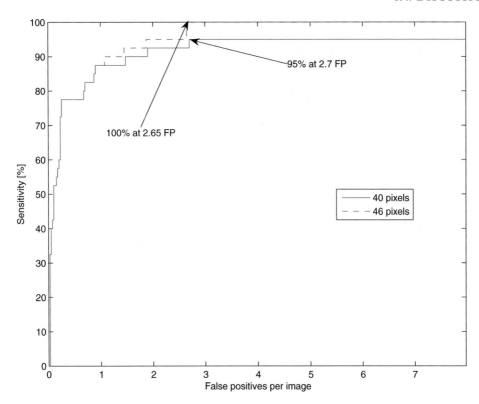

Figure 6.5: FROC curves for the evaluation of detection of the ONH using phase portraits with the DRIVE dataset (40 images). The dashed line corresponds to the FROC curve with a threshold for successful detection of 46 pixels from the manually marked to the detected center. The solid line is the FROC obtained when the same value is set to be 40 pixels. FP = false positive. Reproduced with permission from R.M. Rangayyan, X. Zhu, F.J. Ayres, and A.L. Ells, "Detection of the optic nerve head in fundus images of the retina with Gabor filters and phase portrait analysis", *Journal of Digital Imaging*, 23(4): 438-453, August 2010. © Springer.

Figure 6.7, the peak on or near the macula ranked the first among the peaks in the node map, and could lead to failure in the detection of the ONH if no intensity-based condition is applied. On the other hand, it could also be possible to utilize this feature of the algorithm to locate the macula.

In Figure 6.8, four examples from the STARE dataset are shown where the ONH was not detected by the described method. In parts (a), (b), and (c) of the figure, the reason for failure is vessel bifurcation, leading to a high peak in node map. In the case of the image in part (d) of the same figure, camera noise at the edge of the FOV caused failure.

From Table 6.1, we can observe that, with the inclusion of selection based on the reference intensity, the average distance between the detected and manually marked ONH was reduced, leading

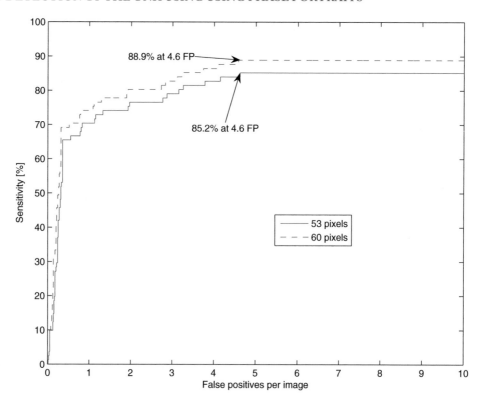

Figure 6.6: FROC curves for the evaluation of detection of the ONH using phase portraits with the STARE dataset (81 images). The dashed line corresponds to the FROC curve with a threshold for successful detection of 60 pixels from the manually marked center to the detected center. The solid line is the FROC obtained when the threshold is set to be 53 pixels. FP = false positive. Reproduced with permission from R.M. Rangayyan, X. Zhu, F.J. Ayres, and A.L. Ells, "Detection of the optic nerve head in fundus images of the retina with Gabor filters and phase portrait analysis", *Journal of Digital Imaging*, 23(4): 438-453, August 2010. © Springer.

to better performance of the described method with the DRIVE images. However, from Table 6.2, we find that the performance in the case of the STARE images with intensity-based selection is the same as that without the condition. This is because the threshold of 50% of the reference intensity is relatively low, and most of the peaks in the node map pass this threshold. The use of a higher threshold could lead to no peak being retained after application of the intensity-based condition in several STARE images where the node map has only one or two peaks.

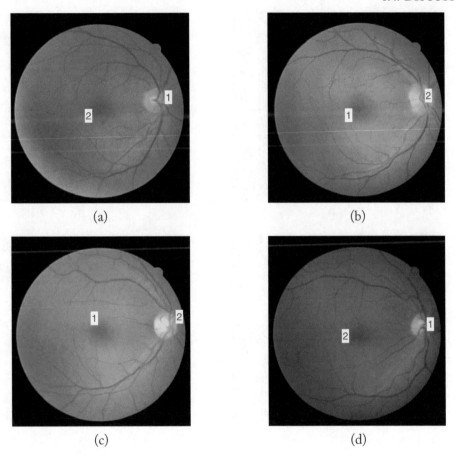

Figure 6.7: (a) Test image 06 from the DRIVE dataset. Each number marked is the rank order of the corresponding peak in the node map. Only the peaks on or near the ONH and macula are shown. (b) Training image 28. (c) Training image 38. (d) Training image 40.

6.4.1   COMPARATIVE ANALYSIS

The two methods described in the present book for the detection of the ONH have their own advantages and disadvantages. From Table 6.3, we see that the method based on phase portraits has success rates of 100% and 69.1% for the DRIVE and the STARE datasets, respectively, which are higher than those provided by the method based on the Hough transform.

By comparing Table 6.1 and Table 5.1, we see that the mean distance related to the results of the method based on phase portraits is larger than that for the method based on the Hough transform (for the DRIVE dataset). The ophthalmologist marked the center of the ONH and not the focal point of convergence of the central retinal vein and artery, which often appear to be different in

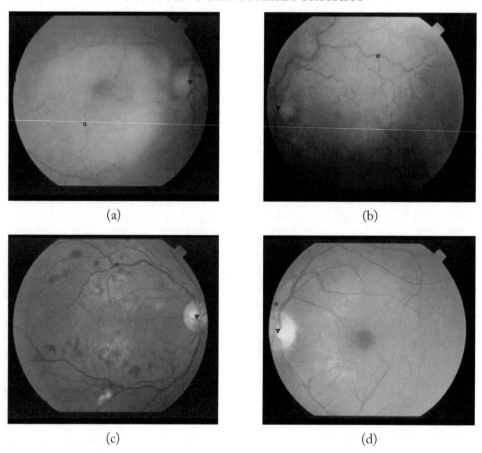

(a)

(b)

(c)

(d)

Figure 6.8: (a) Image im0004 from the STARE dataset. The black square is the first peak detected in the node map that also met the condition based on 50% of the reference intensity. The black triangle is the center of the ONH marked by the ophthalmologist. Distance = 375 pixels. (b) Image im0027. Distance = 368 pixels. (c) Image im0139. Distance = 312 pixels. (d) Image im0239. Distance = 84 pixels. Reproduced with permission from R.M. Rangayyan, X. Zhu, F.J. Ayres, and A.L. Ells, "Detection of the optic nerve head in fundus images of the retina with Gabor filters and phase portrait analysis", *Journal of Digital Imaging*, 23(4): 438-453, August 2010. © Springer.

fundus images of the retina, especially in the case of glaucoma. With erosion of the optic cup due to glaucoma, the central retinal artery and vein are displaced to a side of the ONH. If the focal point of convergence of the central retinal vein and artery is considered as the reference, the mean distance of the results of detection provided by the method using phase portraits could be lower.

Figure 6.9 shows the results of detection of the ONH with four examples from the STARE dataset. The detected circle in part (a) of Figure 6.9 is considered to be a successful detection with

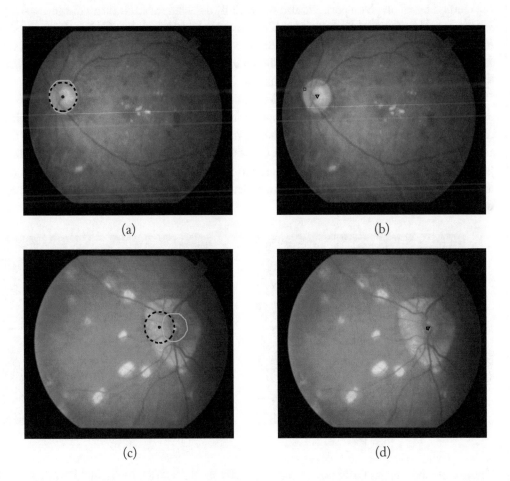

Figure 6.9: (a) The result of detection of the ONH for the image im0009 from the STARE dataset using the Hough transform. The black dashed circle corresponds to the highest local maximum in the Hough space that also meets the condition based on 90% of the reference intensity. The cyan contour in solid line is the contour of the ONH marked by the ophthalmologist. Distance = 3 pixels, overlap = 0.87 with the intensity condition. (b) The result of detection for the image im0009 using phase portraits. The square represents the first peak detected in the node map that also meets the condition based on 50% of the reference intensity. The triangle indicates the center of the ONH marked by the ophthalmologist. Distance = 47 pixels. (c) Results for the STARE image im0010 using the Hough transform. Distance = 51 pixels, overlap = 0.22 with the intensity condition. (d) Results for the image im0010 using phase portraits; distance = 2.2 pixels.

distance = 3 pixels and overlap = 0.87 with the intensity condition and using the Hough transform; the method based on phase portraits also successfully detected the ONH with distance = 47.4 pixels, as shown in part (b) of the same figure. The image im0010 from the STARE dataset is shown in Figure 6.9 (c); in this case, the phase portrait method has detected the center of the ONH more accurately than the Hough transform, as shown in Figure 6.9 (d). The Hough transform was misled by white scar and vessel curvature in the image.

Some of the images in the STARE dataset are out of focus. In Figure 6.10 (a), the result of detection for image im0035 using the Hough transform is shown. Because the image is blurry, it is difficult to detect the edges in the images; hence, the Hough transform could not succeed in detecting an appropriate circle to fit the ONH. Part (b) of the same figure shows the result of detection using phase portraits, which is a successful detection with distance = 26 pixels. The advantage of the method based on phase portraits is that it can detect the vascular pattern existing in the image even when the image is blurry.

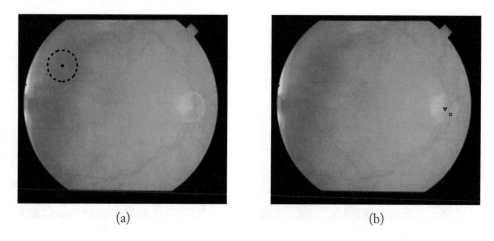

(a) (b)

Figure 6.10: (a) The results of detection of the ONH for the STARE image im0035 using the Hough transform. The black dashed circle corresponds to the highest local maximum in the Hough space that also meets the condition based on 90% of the reference intensity. The cyan contour in solid line is the contour of the ONH marked by the ophthalmologist. Distance = 454 pixels, overlap = 0 with the intensity condition. (b) The result of detection using phase portraits. The black square is the first peak detected in the node map that also meets the condition based on 50% of the reference intensity. The black triangle is the center of the ONH marked by the ophthalmologist. Distance = 26 pixels.

6.5   REMARKS

A procedure based on phase portraits to locate the ONH was described in detail in this chapter; no similar method has been reported in any of the published works on the detection of the ONH. The blood vessels of the retina were detected using Gabor filters and phase portrait modeling was applied to the orientation field to detect points of convergence of the vessels. The method was evaluated by using the distance from the detected center of the ONH to that marked independently by an ophthalmologist. With the inclusion of a step for intensity-based selection of the peaks in the node map, a successful detection rate of 100% was obtained with the 40 images in the DRIVE dataset.

CHAPTER 7

Concluding Remarks

Digital image processing and pattern analysis techniques for the detection of the ONH in fundus images of the retina were described in the book. Two methods based on the Hough transform and phase portraits were described in Chapter 5 and Chapter 6, respectively.

The Hough transform approach is based on the properties of the ONH, including edge detection using the Sobel or the Canny method, and detection of circles using the Hough transform. With an intensity-based condition for the selection of circles and a limit of 46 pixels (0.92 mm) on the distance between the center of the detected circle and the manually identified center of the ONH, a successful detection rate of 90% was achieved with 40 images from the DRIVE dataset. However, the method performed poorly with the images from the STARE dataset, providing a successful detection rate of only 44.4% with 81 images. Analysis of the results was performed in two different ways. One approach involved the assessment of the overlap and distance between the manually identified and detected ONHs, independently marked by an ophthalmologist. The second approach was based on FROC analysis.

The phase portrait method detects points of convergence of vessels, and includes steps for the detection of the blood vessels of the retina using Gabor filters and phase portrait modeling of the orientation field of the image. The method was evaluated by measuring the distance from the detected center to the manually marked center of the ONH. With an intensity-based condition for the selection of peaks in the node map, a successful detection rate of 100% was obtained with the DRIVE images. FROC analysis indicated a sensitivity of 100% at 2.65 false positives per image with the 40 images in the DRIVE dataset. With the 81 images from the STARE dataset, the method provided a successful detection rate of 69.1% and a sensitivity of 88.9% at 4.6 false positives per image.

By comparative analysis of the results of the two methods, we can conclude that the phase portrait method performs better in locating the center of the ONH. However, it could be advantageous to develop a strategy to combine the results of the two methods.

The methods performed well with the DRIVE images but yielded poor results with the STARE images. Further studies are required to incorporate additional characteristics of the ONH to improve the efficiency of detection. Additional constraints may be required for successful detection of the ONH in images with abnormal features and pathological characteristics.

The phase portrait method could be extended for the detection of the macula. The Hough transform could be applied to detect the arcades of the central retinal vein and artery [19, 20].

The methods could be incorporated in a CAD system to assist in the diagnosis and treatment of retinal pathology.

References

[1] Michaelson IC and Benezra D. *Textbook of the Fundus of the Eye*. Churchill Livingstone, Edinburgh, UK, 3rd edition, 1980. 1, 2, 3, 9, 11, 12

[2] Glasspool MG. *Atlas of Ophthalmology*. University Park Press, Baltimore, MD, 1982. 1, 3

[3] Acharya R, Tan W, Yun WL, Ng EYK, Min LC, Chee C, Gupta M, Nayak J, and Suri JS. The human eye. In Acharya R, Ng EYK, and Suri JS, editors, *Image Modeling of the Human Eye*, pages 1–35. Artech House, Norwood, MA, 2008. 1, 6

[4] Swanson C, Cocker KD, Parker KH, Moseley MJ, and Fielder AR. Semiautomated computer analysis of vessel growth in preterm infants without and with ROP. *British Journal of Ophthalmology*, 87(12):1474–1477, 2003. DOI: 10.1136/bjo.87.12.1474 2, 23

[5] Wallace DK, Jomier J, Aylward SR, and Landers, III MB. Computer-automated quantification of Plus disease in retinopathy of prematurity. *Journal of American Association for Pediatric Ophthalmology and Strabismus*, 7:126–130, April 2003. DOI: 10.1016/S1091-8531(02)00015-0 2

[6] Gelman R, Martinez-Perez ME, Vanderveen DK, Moskowitz A, and Fulton AB. Diagnosis of Plus disease in retinopathy of prematurity using retinal image multiscale analysis. *Investigative Ophthalmology & Visual Science*, 46(12):4734–4738, 2005. DOI: 10.1167/iovs.05-0646 2

[7] Ells A, Holmes JM, Astle WF, Williams G, Leske DA, Fielden M, Uphill B, Jennett P, and Hebert M. Telemedicine approach to screening for severe retinopathy of prematurity: A pilot study. *American Academy of Ophthalmology*, 110(11):2113–2117, 2003. DOI: 10.1016/S0161-6420(03)00831-5 2, 6, 12, 13

[8] Staal J, Abràmoff MD, Niemeijer M, Viergever MA, and van Ginneken B. Ridge-based vessel segmentation in color images of the retina. *IEEE Transactions on Medical Imaging*, 23(4):501–509, 2004. DOI: 10.1109/TMI.2004.825627 2, 9, 12, 21, 37

[9] DRIVE: Digital Retinal Images for Vessel Extraction, http:// www. isi.uu.nl/ Research/ Databases/ DRIVE/, accessed on September 30, 2010. 2, 9, 37

[10] Patton N, Aslam TM, MacGillivray T, Deary IJ, Dhillon B, Eikelboom RH, Yogesan K, and Constable IJ. Retinal image analysis: Concepts, applications and potential. *Progress in Retinal and Eye Research*, 25(1):99–127, 2006. DOI: 10.1016/j.preteyeres.2005.07.001 4, 6, 11, 12, 23

[11] Li H and Chutatape O. Automated feature extraction in color retinal images by a model based approach. *IEEE Transactions on Biomedical Engineering*, 51(2):246–254, 2004. DOI: 10.1109/TBME.2003.820400 4, 5, 9, 11, 12

[12] Rangayyan RM, Ayres FJ, Oloumi F, Oloumi F, and Eshghzadeh-Zanjani P. Detection of blood vessels in the retina with multiscale Gabor filters. *Journal of Electronic Imaging*, 17(2):023018:1–7, April-June 2008. DOI: 10.1117/1.2907209 4, 12, 17, 22, 23, 59

[13] Fleming AD, Goatman KA, Philip S, Olson JA, and Sharp PF. Automatic detection of retinal anatomy to assist diabetic retinopathy screening. *Physics in Medicine and Biology*, 52:331–345, 2007. DOI: 10.1088/0031-9155/52/2/002 4, 5, 6, 11, 12, 54, 66

[14] Retinopathy Online Challenge, the University of Iowa, http:// roc. healthcare. uiowa. edu/, accessed on September 16, 2008. 4

[15] Hoover A, Kouznetsova V, and Goldbaum M. Locating blood vessels in retinal images by piecewise threshold probing of a matched filter response. *IEEE Transactions on Medical Imaging*, 19(3):203–210, 2000. DOI: 10.1109/42.845178 4, 21, 37

[16] Hoover A and Goldbaum M. Locating the optic nerve in a retinal image using the fuzzy convergence of the blood vessels. *IEEE Transactions on Medical Imaging*, 22(8):951–958, August 2003. DOI: 10.1109/TMI.2003.815900 4, 10, 12, 37, 38, 41, 54, 65, 66

[17] Structured Analysis of the Retina, http:// www. ces. clemson. edu /~ahoover/ stare/, accessed on March 24, 2008. 4, 37, 38

[18] Tobin KW, Chaum E, Govindasamy VP, and Karnowski TP. Detection of anatomic structures in human retinal imagery. *IEEE Transactions on Medical Imaging*, 26(12):1729–1739, December 2007. DOI: 10.1109/TMI.2007.902801 5, 9, 11, 12

[19] Oloumi F and Rangayyan RM. Detection of the temporal arcade in fundus images of the retina using the Hough transform. In *Engineering in Medicine and Biology Society, Annual International Conference of the IEEE*, pages 3585 –3588, September 2009. DOI: 10.1109/IEMBS.2009.5335389 5, 75

[20] Oloumi F, Rangayyan RM, and Ells AL. Parametric representation of the retinal temporal arcade. In *The 10th IEEE International Conference on Information Technology and Applications in Biomedicine*, Corfu, Greece, November 2010. DOI: 10.1109/ITAB.2010.5687722 5, 75

[21] Early Treatment Diabetic Retinopathy Study Research Group. Grading diabetic retinopathy from stereoscopic color fundus photographs -An extension of the modified Airlie House classification (ETDRS report number 10). *Ophthalmology*, 98:786–806, 1991. 5, 6

[22] Sun H and Nathans J. The challenge of macular degeneration. *Scientific American*, 285(4):68–75, 2001. DOI: 10.1038/scientificamerican1001-68 6

[23] Sinthanayothin C, Boyce JF, Cook HL, and Williamson TH. Automated localisation of the optic disc, fovea, and retinal blood vessels from digital colour fundus images. *British Journal of Ophthalmology*, 83(4):902–910, August 1999. DOI: 10.1136/bjo.83.8.902 9, 11, 12, 54, 66

[24] Niemeijer M, Abràmoff MD, and van Ginneken B. Segmentation of the optic disk, macula and vascular arch in fundus photographs. *IEEE Transactions on Medical Imaging*, 26(1):116–127, 2007. DOI: 10.1109/TMI.2006.885336 9

[25] Park M, Jin JS, and Luo S. Locating the optic disc in retinal images. In *Proceedings of the International Conference on Computer Graphics, Imaging and Visualisation*, page 5, Sydney, Qld., Australia, July, 2006. IEEE. DOI: 10.1109/CGIV.2006.63 9, 54, 56, 66

[26] Barrett SF, Naess E, and Molvik T. Employing the Hough transform to locate the optic disk. *Biomedical Sciences Instrumentation*, 37:81–86, 2001. 9, 53

[27] ter Haar F. Automatic localization of the optic disc in digital colour images of the human retina. Master's thesis, Utrecht University, Utrecht, the Netherlands, 2005. 9, 10, 38, 40, 41, 46, 53, 54, 66

[28] Chrástek R, Skokan M, Kubecka L, Wolf M, Donath K, Jan J, Michelson G, and Niemann H. Multimodal retinal image registration for optic disk segmentation. In *Methods of Information in Medicine*, volume 43, pages 336–42, 2004. 9, 37, 53

[29] Chrástek R, Wolf M, Donath K, Niemann H, Paulus D, Hothorn T, Lausen B, Lämmer R, Mardin CY, and Michelson G. Automated segmentation of the optic nerve head for diagnosis of glaucoma. *Medical Image Analysis*, 9(4):297–314, 2005. DOI: 10.1016/j.media.2004.12.004 9, 53

[30] Osareh A, Mirmehd M, Thomas B, and Markham R. Comparison of colour spaces for optic disc localisation in retinal images. In *Proceedings 16th International Conference on Pattern Recognition*, pages 743–746, Quebec City, Quebec, Canada, 2002. DOI: 10.1109/ICPR.2002.1044865 9, 54, 66

[31] Youssif AAHAR, Ghalwash AZ, and Ghoneim AASAR. Optic disc detection from normalized digital fundus images by means of a vessels' direction matched filter. *IEEE Transactions on Medical Imaging*, 27(1):11–18, 2008. DOI: 10.1109/TMI.2007.900326 10, 41, 52, 54, 56, 65, 66

[32] Lalonde M, Beaulieu M, and Gagnon L. Fast and robust optic disc detection using pyramidal decomposition and Hausdorff-based template matching. *IEEE Transactions on Medical Imaging*, 20(11):1193–1200, 2001. DOI: 10.1109/42.963823 10, 46, 54, 66

[33] Foracchia M, Grisan E, and Ruggeri A. Detection of optic disc in retinal images by means of a geometrical model of vessel structure. *IEEE Transactions on Medical Imaging*, 23(10):1189–1195, 2004. DOI: 10.1109/TMI.2004.829331 10, 12, 54, 66

[34] Ying H, Zhang M, and Liu JC. Fractal-based automatic localization and segmentation of optic disc in retinal images. In *Proceedings of the 29th Annual International Conference of the IEEE Engineering in Medicine and Biology Society*, pages 4139–4141, Lyon, France, August 23-26, 2007. IEEE. DOI: 10.1109/IEMBS.2007.4353247 10, 54, 56, 66

[35] Kim SK, Kong HJ, Seo JM, Cho BJ, Park KH, Hwang JM, Kim DM, Chung H, and Kim HC. Segmentation of optic nerve head using warping and RANSAC. In *Proceedings of the 29th Annual International Conference of the IEEE Engineering in Medicine and Biology Society*, pages 900–903, Lyon, France, August 23-26, 2007. IEEE. DOI: 10.1109/IEMBS.2007.4352436 10, 54, 66

[36] Park J, Kien NT, and Lee G. Optic disc detection in retinal images using tensor voting and adaptive mean-shift. In *IEEE 3rd International Conference on Intelligent Computer Communication and Processing, ICCP*, pages 237–241, Cluj-Napoca, Romania, 2007. DOI: 10.1109/ICCP.2007.4352167 11, 54, 66

[37] Carmona EJ, Rincon M, García-Feijoó J, and Martínez de-la Casa JM. Identification of the optic nerve head with genetic algorithms. *Artificial Intelligence in Medicine*, 43(3):243–59, July 2008. DOI: 10.1016/j.artmed.2008.04.005 11

[38] Hussain AR. Optic nerve head segmentation using genetic active contours. In *Proceeding International Conference on Computer and Communication Engineering*, pages 783–787, Kuala Lumpur, Malaysia, May 13-15, 2008. IEEE. DOI: 10.1109/ICCCE.2008.4580712 11

[39] Pinz A, Bernogger S, Datlinger P, and Kruger A. Mapping the human retina. *IEEE Transactions on Medical Imaging*, 17(4):606 – 619, 1998. DOI: 10.1109/42.730405 11, 12

[40] Chaudhuri S, Chatterjee S, Katz N, Nelson M, and Goldbaum M. Detection of blood vessels in retinal images using two-dimensional matched filters. *IEEE Transactions on Medical Imaging*, 8(3):263–269, 1989. DOI: 10.1109/42.34715 12, 21

[41] Narasimha-Iyer H, Can A, Roysam B, Stewart CV, Tanenbaum HL, Majerovics A, and Singh H. Robust detection and classification of longitudinal changes in color retinal fundus images for monitoring diabetic retinopathy. *IEEE Transactions on Biomedical Engineering*, 53(6):1084–1098, 2006. DOI: 10.1109/TBME.2005.863971 12

[42] Lowell J, Hunter A, Steel D, Basu A, Ryder R, and Kennedy RL. Measurement of retinal vessel widths from fundus images based on 2-D modeling. *IEEE Transactions on Medical Imaging*, 23(10):1196–1204, 2004. DOI: 10.1109/TMI.2004.830524 12

[43] Soares JVB, Leandro JJG, Cesar Jr. RM, Jelinek HF, and Cree MJ. Retinal vessel segmentation using the 2-D Gabor wavelet and supervised classification. *IEEE Transactions on Medical Imaging*, 25(9):1214–1222, 2006. DOI: 10.1109/TMI.2006.879967 12, 17

[44] Kochner B, Schuhmann D, Michaelis M, Mann G, and Englmeier KH. Course tracking and contour extraction of retinal vessels from color fundus photographs: most efficient use of steerable filters for model based image analysis. In *Proceedings of the SPIE, The International Society for Optical Engineering*, pages 755–761, San Diego, CA, February, 1998. SPIE. DOI: 10.1117/12.310955 12

[45] Gregson PH, Shen Z, Scott RC, and Kozousek V. Automated grading of venous beading. *Computers and Biomedical Research*, 28:291–304, August 1995. DOI: 10.1006/cbmr.1995.1020 12

[46] Zana F and Klein JC. Segmentation of vessel-like patterns using mathematical morphology and curvature estimation. *IEEE Transactions on Image Processing*, 10(7):1010–1019, July 2001. DOI: 10.1109/83.931095 12

[47] Jiang X and Mojon D. Adaptive local thresholding by verification-based multithreshold probing with application to vessel detection in retinal images. *IEEE Transactions on Pattern Analysis and Machine Intelligence*, 25(1):131–137, 2003. DOI: 10.1109/TPAMI.2003.1159954 12

[48] Gang L, Chutatape O, and Krishnan SM. Detection and measurement of retinal vessels in fundus images using amplitude modified second-order Gaussian filter. *IEEE Transactions on Biomedical Engineering*, 49(2):168–172, 2002. DOI: 10.1109/10.979356 12

[49] Matsopoulos GK, Asvestas PA, Delibasis KK, Mouravliansky NA, and Zeyen TG. Detection of glaucomatous change based on vessel shape analysis. *Computerized Medical Imaging and Graphics*, 32(3):183–192, April 2008. DOI: 10.1016/j.compmedimag.2007.11.003 12

[50] Niemeijer M, Ginneken BV, Staal J, Suttorp-Schulten MS, and Abramoff MD. Automatic detection of red lesions in digital color fundus photographs. *IEEE Transactions on Medical Imaging*, 24(5):584–592, May 2005. DOI: 10.1109/TMI.2005.843738 12

[51] Hatanaka Y, Nakagawa T, Hayashi Y, Hara T, and Fujita H. Improvement of automated detection method of hemorrhages in fundus images. In *Proceedings of the 30th Annual International Conference of the IEEE Engineering in Medicine and Biology Society*, pages 5429–5432, Vancouver, BC, Canada, August, 2008. IEEE. DOI: 10.1109/IEMBS.2008.4650442 12

[52] Liu Z, Opas C, and Krishnan SM. Automatic image analysis of fundus photograph. In *Proceedings of the 19th Annual International Conference of the IEEE Engineering in Medicine and Biology Society*, pages 524–525, Chicago, IL, October 1997. IEEE. DOI: 10.1109/IEMBS.1997.757661 12

[53] Wang H, Hsu W, Guan K, and Lee M. An effective approach to detect lesions in color retinal images. In *Proceedings of the IEEE Conference on Computer Vision and Pattern Recognition*, pages 181–186, Hilton Head Island, SC, 2000. IEEE. DOI: 10.1109/CVPR.2000.854775 12

[54] Gonzalez RC and Woods RE. *Digital Image Processing*. Prentice Hall, Upper Saddle River, NJ, 2nd edition, 2002. 15, 17, 24

[55] Rangayyan RM. *Biomedical Image Analysis*. CRC, Boca Raton, FL, 2005. 15, 17, 19, 22, 24, 59

[56] Russ JC. *The Image Processing Handbook*. CRC, Boca Raton, FL, 2nd edition, 1995. 15

[57] Pratt WK. *Digital Image Processing*. John Wiley & Sons, New York, NY, 1978. 17

[58] The MathWorks, http:// www. mathworks. com/, accessed on March 24, 2008. 20, 21, 43

[59] Canny J. A computational approach to edge detection. *IEEE Transactions on Pattern Analysis and Machine Intelligence*, PAMI-8(6):670–698, 1986. DOI: 10.1109/TPAMI.1986.4767851 19

[60] Al-Rawi M, Qutaishat M, and Arrar M. An improved matched filter for blood vessel detection of digital retinal images. *Computers in Biology and Medicine*, 37(2):262–267, February 2007. DOI: 10.1016/j.compbiomed.2006.03.003 21

[61] Rangayyan RM and Ayres FJ. Gabor filters and phase portraits for the detection of architectural distortion in mammograms. *Medical and Biological Engineering and Computing*, 44(10):883–894, October 2006. DOI: 10.1007/s11517-006-0109-2 22, 59

[62] Ayres FJ and Rangayyan RM. Design and performance analysis of oriented feature detectors. *Journal of Electronic Imaging*, 16(2):023007:1–12, 2007. DOI: 10.1117/1.2728751 22, 31, 59

[63] Manjunath BS and Ma WY. Texture features for browsing and retrieval of image data. *IEEE Transactions on Pattern Analysis and Machine Intelligence*, 18(8):837–842, 1996. DOI: 10.1109/34.531803 22, 59

[64] Hough PVC. Method and means for recognizing complex patterns. US Patent 3, 069, 654, December 18, 1962. 24

[65] Duda RO and Hart PE. Use of the Hough transformation to detect lines and curves in pictures. *Communications of the ACM*, 15(1):11–15, January 1972. DOI: 10.1145/361237.361242 24

[66] Rao AR and Jain RC. Computerized flow field analysis: Oriented texture fields. *IEEE Transactions on Pattern Analysis and Machine Intelligence*, 14(7):693–709, 1992. DOI: 10.1109/34.142908 29, 31, 34, 59

[67] Ayres FJ and Rangayyan RM. Reduction of false positives in the detection of architectural distortion in mammograms by using a geometrically constrained phase portrait model. *International Journal of Computer Assisted Radiology and Surgery*, 1(6):361–369, 2007. DOI: 10.1007/s11548-007-0072-x 29, 32, 33, 34, 59, 60

[68] Wylie CR and Barrett LC. *Advanced Engineering Mathematics*. McGraw-Hill, New York, NY, 6th edition, 1995. 29, 32

[69] Image Processing and Analysis in Java, http:// rsbweb. nih.gov/ ij/, accessed on September 3, 2008. 39

[70] Egan JP, Greenberg GZ, and Schulman AI. Operating characteristics, signal detectability, and the method of free response. *The Journal of the Acoustical Society of America*, 33(8):993–1007, 1961. DOI: 10.1121/1.1908935 40

[71] Edwards DC, Kupinski MA, Metz CE, and Nishikawa RM. Maximum likelihood fitting of FROC curves under an initial-detection-and-candidate-analysis model. *Medical Physics*, 29:2861–2870, December 2002. DOI: 10.1118/1.1524631 40

[72] Zhu X, Rangayyan RM, and Ells AL. Detection of the optic disc in fundus images of the retina using the Hough transform for circles. *Journal of Digital Imaging*, 23(3):332–341, June 2010. DOI: 10.1007/s10278-009-9189-5 43, 54, 56, 66

[73] Zhu X and Rangayyan RM. Detection of the optic disc in images of the retina using the Hough transform. In *Proceedings of the 30th Annual International Conference of the IEEE Engineering in Medicine and Biology Society*, pages 3546–3549, Vancouver, BC, Canada, August 20-24, 2008. IEEE. DOI: 10.1109/IEMBS.2008.4649971 55

[74] Rangayyan RM, Zhu X, Ayres FJ, and Ells AL. Detection of the optic nerve head in fundus images of the retina with Gabor filters and phase portrait analysis. *Journal of Digital Imaging*, 23(4):438–453, August 2010. DOI: 10.1007/s10278-009-9261-1 59, 66

Authors' Biographies

XIAOLU (IRIS) ZHU

Xiaolu (Iris) Zhu received the Bachelor of Engineering degree in Electrical Engineering in 2006 from Beijing University of Posts and Telecommunications, Beijing, P.R. China. During the peorid of 2007-2008, she obtained the Master of Science degree in Electrical Engineering from the University of Calgary, Alberta, Canada. She has published three papers in journals and three in proceedings of conferences. At present, she is working as a software developer in the Institute for Biodiagnostics (West), National Research Council of Canada, Calgary, Alberta, Canada. Her research interests are digital signal and image processing as well as biomedical imaging. In her spare time, she loves to swim, draw, and travel.

RANGARAJ M. RANGAYYAN

Rangaraj M. Rangayyan is a Professor with the Department of Electrical and Computer Engineering, and an Adjunct Professor of Surgery and Radiology, at the University of Calgary, Calgary, Alberta, Canada. He received the Bachelor of Engineering degree in Electronics and Communication in 1976 from the University of Mysore at the People's Education Society College of Engineering, Mandya, Karnataka, India, and the Ph.D. degree in Electrical Engineering from the Indian Institute of Science, Bangalore, Karnataka, India, in 1980. His research interests are in the areas of digital signal and image processing, biomedical signal analysis, biomedical image analysis, and computer-aided diagnosis. He has published more than 140 papers in journals and 220 papers in proceedings of conferences. His research productivity was recognized with the 1997 and 2001 Research Excellence Awards of the Department of Electrical and Computer Engineering, the 1997 Research Award of the Faculty of Engineering, and by appointment as a "University Professor" in 2003, at the University of Calgary. He is the author of two textbooks: *Biomedical Signal Analysis* (IEEE/ Wiley, 2002) and *Biomedical Image Analysis* (CRC, 2005); he has coauthored and coedited several other books. He was recognized by the IEEE with the award of the Third Millennium Medal in 2000, and he was elected as a Fellow of the IEEE in 2001, Fellow of the Engineering Institute of Canada in 2002, Fellow of the American Institute for Medical and Biological Engineering in 2003, Fellow of SPIE: the International Society for Optical Engineering in 2003, Fellow of the Society for Imaging Informatics in Medicine in 2007, Fellow of the Canadian Medical and Biological Engineering Society in 2007, and Fellow of the Canadian Academy of Engineering in 2009. He has been awarded the Killam Resident Fellowship thrice (1998, 2002, and 2007) in support of his book-writing projects.

ANNA L. ELLS

Dr. Anna L. Ells is an ophthalmologist, with dual fellowships in Pediatric Ophthalmology and Medical Retina. She has a combined academic hospital-based practice and private practice. Dr. Ells's research focuses on retinopathy of prematurity (ROP), global prevention of blindness in children, and telemedicine approaches to ROP. Dr. Ells has international expertise and has published extensively in peer-reviewed journals.

Index

Printed in the United States
by Baker & Taylor Publisher Services